CREATION OF UNIVERSE
GOD
OR BIG BANG

CREATION OF UNIVERSE
GOD
OR BIG BANG

PROF (LT COL) KEERTHI KUMAR PATANGAY

Notion Press

Old No. 38, New No. 6
McNichols Road, Chetpet
Chennai - 600 031

First Published by Notion Press 2017
Copyright © Prof (Lt Col) Keerthi Kumar Patangay 2017
All Rights Reserved.

ISBN 978-1-947988-39-2

This book has been published with all reasonable efforts taken to make the material error-free after the consent of the author. No part of this book shall be used, reproduced in any manner whatsoever without written permission from the author, except in the case of brief quotations embodied in critical articles and reviews.

The Author of this book is solely responsible and liable for its content including but not limited to the views, representations, descriptions, statements, information, opinions and references ["Content"]. The Content of this book shall not constitute or be construed or deemed to reflect the opinion or expression of the Publisher or Editor. Neither the Publisher nor Editor endorse or approve the Content of this book or guarantee the reliability, accuracy or completeness of the Content published herein and do not make any representations or warranties of any kind, express or implied, including but not limited to the implied warranties of merchantability, fitness for a particular purpose. The Publisher and Editor shall not be liable whatsoever for any errors, omissions, whether such errors or omissions result from negligence, accident, or any other cause or claims for loss or damages of any kind, including without limitation, indirect or consequential loss or damage arising out of use, inability to use, or about the reliability, accuracy or sufficiency of the information contained in this book.

DEDICATED TO MY SISTER SANT ANUSUYA BAI PATANGAY SHORT BIOGRAPHY

Late Sant Anusuya Bai Patangay
23rd July 1937 to 17 Oct 1975

We are three sisters and three brothers. All threesisters were elder to us. Sant Anusuya Bai Patangay developed Tuberculosis of Spine when she was 5 years old. At that time antitubercular treatment in India was not available. By the time Streptomycin came into market she had already developed gibbusat her back. A surgeon who came to Hyderabad from England took the challenge to treat her. He put her on plaster of Paris cast for 6 months along with antitubercular treatment available at that time. When plaster cast was removed she became paraplegic. One widow Brahmin who had lost her husband at very young age started teaching Bhagwat Gita (Bha.G). In no time she memorised whole of Bhagwat

Gita. Then she started to learn Hindi Literature and earned highest degree from Hindi Prachar Sabha, not content with this she started to learn class 10 with sole idea of learning English and she qualified in class 10 with first class. Thereafter she was teaching Bha.G to young girls in late 1960s. Then she started to visit Brindavan, Mathura and Kashi delivering sermons on Bha.G. My niece was always attending her. When she came to know that her end has come she quietly left our home and went to Kashi where she attained Moksha (salvation).

CONTENTS

Foreword . . . . *ix*
Preface . . . . *xi*
Acknowledgements . . . . *xiii*
Prayers on Impersonal God . . . . *xv*
Some Common Abbreviations . . . . *xix*

CHAPTER 1:   Who Created this Universe? 1
CHAPTER 2:   Astronomy 4
CHAPTER 3:   Pramana 14
CHAPTER 4:   Sāmkhya/Sānkhya Sutra Kapila. 25
CHAPTER 5:   Creation as per Rig Veda 68
CHAPTER 6:   The Creation in Rig Veda X:129 83
CHAPTER 7:   The Creation of the Universe –
             The Brihadaranyaka Upanishad 107
CHAPTER 8:   Aitareya Upanishad. 141
CHAPTER 9:   What Is the Source of Creation
             According to Upanishads? 154
CHAPTER 10: Manohar Patil. Srimad Bhagwatam is my life 163
CHAPTER 11: Manvantara Theory of Evolution of Solar System . . 175
CHAPTER 12: Hindu Units of Time. 185

CHAPTER 13: Speed of Light in Rig Veda 196

CHAPTER 14: Where Do We Stand in Universe 199

CHAPTER 15: The Biblical Doctrine of Creation 207

CHAPTER 16: Creation as per Islam-Holy Quran 234

CHAPTER 17: Big Bang 247

CHAPTER 18: No Big Bang? Quantum Equation Predicts Universe Has No Beginning. 275

CHAPTER 19: Top Ten Mind Bending Theories – TopTenz.net 279

CHAPTER 20: A Universe from Nothing? 289

CHAPTER 21: Multiverse (Religion) 295

CHAPTER 22: Is the Universe Conscious?. 298

CHAPTER 23: Conclusion 302

FOREWORD

I am extremely happy to mwrite a brief forword to the interesting book by Lt Col Keerthi Kumar Patangay. The book takes the subject of creation of universe. There are two quite different approaches. One is based on several traditional Hindu such as Rig Veda, Upanishads, darshanas such as Sankhya and other. Alternatively there is modern scientific conjecture that the universe began with big bang. Ofcourse this view has drawn criticism from several eminent scieritsts also. The Book with its 22 chapters devotes about eleven chapters for views of Hindu scripture & 2 chapers for views from Bible and Quran, and the remaining eight chapters to the scientific view such as big bang abd related topics.

The author is very modest but has great shcolorship. While discussing creation ideas from Hindu Scriptures, he gives extensive qoutations from scholars and savants who have extensive knowlwdge such as Pujya Krishnananda, on Brihadaranyaka Upanishad, Seetha Madhava Aitereya Upanishad, Manvantara Theory of Evolution of Solar Universe by Dhani, Rig Veda by RL Kashyap ect,. The author expresses his own views on Hindu ideas of time, Shiva, Vishnu and others supported by extensive quotations from the Upanishadas, Vishnu and Shiva Purana. His treatment of scientific view is also fair. He discusses the theory based on big bang, those not involving big bang and the ten so called mind bending theories.

I only added a few words on chapter 22 titled, "Is Universe Conscious", taken by a book by Steve Steward-Williams, titled "Darwin God and

Foreword

Meaninf of Life." Many Western writers like this author and their Indian followers writing from the plane of mental ideas whose limitations have been documented by several Western psychologists. Author poses simple question,

"Is piece of stone conscious." His answer is typical no. He does not know even modern physics that when a person taken a stone in his hand, there is bound to br interaction. If a person is very conscious he feels consciousness in stone. An ordinary person does not do so. All these ideas are explained simply and beautifully in conversational style in the Indian languages, such as Kanada, Telugu ect. Why did the Cern, the major centre for study on particle physics instill a beautiful statue of the deva Nataraja, the dancing Shiva? Why do atleast the 50 thousand persons, every visit the temple of Lord Srinivasa on the hill in Tirpati in Andhra Pradesh? Are they are stupid? The visitors include very famous engineers, scientists, medical professionls "such as Dr. Reddy who started the chain of Apollo Hospitals" and highly creative persons. Here the consciousness of the visitors is in touch with the consciousness of the murthy or statue (7 feet tall).

The author of the book is himself highly qualified medical professional should be congratulated on bringing out such an excellent book. I invoke blessings of the Devine Mother for his well being and creativity.

R. L. Kashyap
July 15, 2017
SAKSHI,
Bengaluru, India.

PREFACE

Denied a translated copy of Rig Veda by fellow Surgeon at Princess Esra Hospital, Hyderabad for the simple reason that I am nota Brahmin made me search for Rig Veda. I could not find Rig Veda in Google search. Then I went to Ramakrishna mutt. There also I could not get Rig Veda (RV) then I collected all the Major Upanishads numbering ten. Then I started reading Upanishads. In every Upanishad that I read there was Creation of Universe as content. Mean while my class fellow in Florida sent me a link on translated RV written by Ralph TH Grifith. I found the English translation of RV by Raiph TH Grifith. He had copied from Sayana, a minister in Vijayanagara dynasty in 17th century AD and also from Max Muller's translation. I found RV tanslated into Englidh by Ralph TH Grifith uninspiring becase it only gives translation but lacks in spirituality. Again my search for RV began. Finally I found Rig Veda along with other vedas in its translation in English by Pujya Sri Dr. RL Kashyap. Unlike Grifith's translation which is translated word to word using Sanskrit-English Dictionary, Pujya Sri Dr. RL Kahyap gives spiritual meaning of Rig Veda. Rig Veda also contain encrypted secret message which was orally transmitted from Guru to his desciples. The secret message is lost and lot of work is being done now by NASA to understand the secret message. NASA found Sanskrit tobe an ideal language for computer programming. Pujya Dr. RL Kashyap has also given the secret message and made reading of all the vedas interesting… However now many scientists are working to interpret the RV. The

RV was traditionallybeing trasmitted to the students orally. Written RV and other vedas started appearing much later, unfortunately much of Vedas has either been either stolen or being lost from generatin to generation in its oral rendition practice. Author does not take credit what soever for the incorporated chapters. Knowledge of Bhagwat Gita, Major Upanishads and Quantum physics facilitates the understanding of last Chapter 'Conclusion.' While discussing quantum physics and Upanishadsin last chapter reader may find it hodgepodge.

ACKNOWLEDGEMENTS

My special thanks go to Sri Pujya RL Kashyap for giving permission to incorporate X:129 of Rig veda on Creation og Universe. I am grateful to Sri Pujya Krishanandji for allowing me to incorporating Cration of Universe-Brihadaranyaka Upanishad. My thanks also go to Rev Dr. G. Wright Doyle for his contribution on Biblical Doctrine on Creation. I am also grateful Dr. (Mrs) Menaka Thadani who undertook EDP work. Last but not the least I am grateful to Dr. Tulsi Ram in Dallas, USA for partly financing this project.

PRAYERS ON IMPERSONAL GOD

ॐ सह नौ अवतु । सह नौ भुनक्तु ।
सह वीर्यम् करवावहै । तेजस्वि नौ अधीतम् अस्तु ।
मा विद्विषणहै । ॐ शान्ति: शान्ति: शान्ति:

Kathopanishad

O God, the Almighty, bless us both (the teacher and the pupil) together, develop us both together, give us strength together. Let the knowledge acquired by us be bright and illumed and second to none. Let both of us live with love, affection and harmony. O God, let there be physical, mental and spirited peace. Om peace, peace, peace.

ॐ पूर्णमद: पूर्णमिदं पूर्णात्पूर्णमुदच्यते ।
पूर्णस्य पूर्णमादाय पूर्णमेवावशिष्यते ॥
ॐ शान्ति: शान्ति: शान्ति: ।

Om Purnamadah purnamidam purnatpurnamudacyate;
Purnasya purnamadaya purnamevavasisyate.
Om santih, santih, santih —

Om. That (Brahman) is infinite, and this (universe) is infinite. The infinite proceeds from the infinite. (Then) taking the infinite (universe), It remains as the infinite (Brahman) alone. Om peace, pease pease.

The other explanation

"This invisible (Brahman) is the full; the visible (the universe) too is full. From the full (Brahman), the full (the visible universe) has come. The full (Brahman) remains same, even after the full (the visible universe) has come out of the full (Brahman)."

ॐ ईशावास्यमिदं सर्वं
यत्किंच जगत्यां जगत् ।
तेन त्यक्तेन न भुञ्जीथा
मा गृधः कस्य स्विद्धनम् ।

Isa Upanishad

**Om isavasyaamidam sarvam yaktinca jagatam jagat
Tena tyaktena bhunjitha ma gradhah kasya sviddhananam.**

In this changing world everything is subject to change, yet everything is covered by the Brahman. Practise renunciation and be strong in consciousness of self. Do not covet i.e., do not run after other's wealth.

नमोनमस्ते- खिन्नकारणाय निष्कारणायाटु भतकारणाय
सम्वागममर्मायसहर्णवाय नमो-पवर्गाय परायणाय ॥

15

Gagendramoksha   **Srimad Bhagwatam**

You are the cause of all causes, but you yourself are not the effect of any cause.

You are wonderful cause but the ordinary causes undergo change when they become the effects; but you produce the world with or without undergoing any change.

You are the scriptures and you are the ocean into which all scriptures flow.

You are the bliss of salvation and the refuge of great souls.

I bow down to you.

Prayers on Impersonal God

All things are made through Him, and without Him nothing was made that was made.

In Him was life and the life was the light of men.

And all light shines in darkness and darkness did not comprehend it.

(John Chapter 1 Verse 3-4)

न नित्यशुच्डबुच्डमुत्कस्वभावस्य तघो
गस्तघो-
गाह्ते

Samkhya Sutra 19

Aph. 19 of Samkhya sutra "In the perpetual purity means the being ever devoid of merit and demerit; the perpetual intelligence means the consisting of uninterrupted thought; and the perpetual liberties means the being dissociated from real pain: that is to say, the connexion with pain in shape of reflection is not a real bondage, (any more than the reflection of China rose is a real stain in the crystal)."

Explanation: China-rose has stain in the crystal itself which cannot be removed, hence perpetual. A dyed cloth stain is not perpetual because stain can be removed. Similarly pain is not perpetual.

Every thing in Universe is within you,
Ask all from yourself — Rumi.

"Why should I seek? I am same as He. His essence speaks through me. I have been looking for Myself" — Rumi.

All things are made through Him and without him nothing was made.

In Him was lifewas the light of men.

And the light shines in darkness, and the darkness did not comprend it.

(John 3,4,5)

SOME COMMON ABBREVIATIONS

RV	Rig Veda
Svet U	Svetaswetara Upanishad
TA	Taittiriya Aranyka
Tai U	Taittiriya Upanishad
Bha G	Bhagwat Gita
Br. Up	Brihadaranya Upanishad
Ba Su	Brahma Sutra
Cha U	Chandogya Upanishad
Is Up	Isa Upanishad
Ka Up	Katha Upanishad
Ke U	Kena Upanishad
Ma U	Mandukya Upanishad
Sha Br	Satapata Brahmana
SB	Srimad Bhagwatam
BB	Big Bang
Aph	Aphorisms
Sam kar Aph	Samkhya Karika Aphorism
Sam Su Aph	Samkhya Sutra Aphorism
Ait Up	Aitereya Upanishad
Br Sut	Brahma Sutra

CHAPTER 1

WHO CREATED THIS UNIVERSE?

"Understanding the question is half an answer" – Socrates. Ask anyone as to who has created this universe. The answer would be simple and straight forward 'Of course God created this Universe.' Next you ask them 'how do you know that God created this Universe.' The answer would be 'Of course it is written in our Holy Books that God Create this Universe.' Religions of Book (Abrahimic Religion) such as Torah, Holy Bible and Holy Quran are vehement in saying so. It is unquestionable because it is revealed to them by none other than God and that they are Words of God. When we refer these Books we mean Holy Torah, Holy Bible and the Holy Quran which say that God created this Universe. Each of these Books has their respective Personal Gods. In Torah God is referred as Yahweh. In Holy Bible He reveals Himself as 'I Am Who I AM' (Exodus 3-14). Christians say that Jesus is the God for Jesus said "There fore I said that you will die if you don't believe I AM, you will die in sin" (John 8:24). Chrisians also believe that Jesus madethe world "Jesus made all things" (John 1:3) Holy Quran says that Allah created this Universe These Books of Religion are dogmatic; in the sense they are 'Revealed by none other than God, hence unquestionable. When it comes to Quran nothing can be added or subtracted for it is a word of Allah. For the scientist it is Big Bang that was the cause of Creation of Universe.

As per Vedas and Upanishads Brahman (God) exists in Unmanifested form – Nirguna Brahmana and in Manifested form as the Universe – Saguna Brahmana.

When we talk of vedantic philosophy of Sanatana Dharma, it encourages one to question unlike Abrahamic religions. It has a place for atheists, theists, agonists, rationalists, materialists and also to logicians. Incidently logicians when they argue, they never come to any conclusion. The beautiful verse from – Bhagwat Gita (Bhag. G) Says:

सर्गाणामादिरन्तश्च मध्यं चैवाहमर्जुन ।
अध्यात्मविद्या विद्यानां वादः प्रवदतामहम् ॥

Chapter 10:32

In this Lord Krishna says 'among the logicians I am the conclusion.'

Vedic philosophy teaches us to be fearless. It is very difficult to live without fear. Fear is a part of life, e.g. fear of death, fear of future, fear of children's future, fear of sin, fear of God, fear of unknown, fear of death so and so forth. In Vedanta it is said that it is fearful to be fearless, but one who is emancipated in Vedanta is fearless.

Lord Buddha has put the meaning of Fear in the two following sentences:

a) Forget everything and run if you are fearful.

b) Face everything and rise (that is when you are devoid of fear)

If you have faith in God you become fearless. Taittareya Upanishad book II Chapter 4 "One is not subjected to fear at anytime if one knows the Bliss that is Brahman"(GOD).

Holy Bible also tells, "For God has not given us a spirit of fear but of power and love of a sound mind" (Timothy 1:7) "I sought the Lord and He heard me, and delivered me from all fear." (Psalms 34.4)

Dealing with the subject of Creation of Universe is very challenging. Even before scientific evidence started emerging Religious book had already dealt with the subject on Creation of Universe as a plan of God. The subject of Creation of Unverse can be dealt with from various Holy Books/Scientific evidence and also rely on print media/social media and internet. No one was there to witness when creation had taken place and all the more reason one has to heavily depend on the subject matter available in Scripture/ Scientific proof. In simple terms we need evidence either from religious books or from ongoing scientific research. In the Chapter 3 we deal with various forms of Evidences. In Sanskrit evidence/ proof is called Pramana.

CHAPTER 2

ASTRONOMY

Ever since man first looked up at the stars he must have asked these questions "what are the things that are twinkling in the sky and how vast is the sky." He also began to ask "are we alone?" Study of the sky and cosmos is called astronomy. The word astronomy is derived from Latin 'astronomia' which literally means star arrangement i.e., from astron "Star" and "namos," and nemesis "to deal with."

The history of astronomy has shapped the course of human society, connecting science with inbuilt human spirituality. Astronomy helped farmers in carrying out agricultural work which was very useful. From Bible we can learn that three wise men from East were guided by stars when Jesus when was born, (Mathew 2:3.) Halye's comet was interpreted as impending doom.

Earliest to talk on Solar System was from Rig Veda. First they talked about gravity and planetary motion. Earliest mention of Gravity and Solar system can be found in Rig Veda. The Sanskrit language of Rig Veda (RV) is quite different from other vedas which were also written in Sanskrit. Rig Veda gives coded messages. Many Indologists translated RV in English. But the translations were only in letter and not in spirit. They could not understand psyche of Sanatana Dharma (SD). Rig Veda shlokas also give spiritual message and also encrypted secret message only revealed to very few of the pupils of that time. All four vedas were transmitted from generation to generation verbally. Only in 300 BCE or so they were found in written form. Many of the Veda sections are lost or stolen away over a period of time.

RIG VEDA ON GRAVITY

सविता यन्त्रै: पर्थिवीमरम्णादस्कम्भने सविता दयामद्रंहत् ।
अश्वमिवाधुक्षद् धुनिमन्तरिक्षमतूर्तेवदधं सविता समुद्रम ॥

Rig Veda 10.145.1

Which means the sun has tied Earth and other planets through attraction and moves? Them itself as if a trainer moves newly trainees around itself holding their reins.

ते हर्यता हरी वादधाते दिवे-दिवे
देत ते विश्वा भुवनानि येमिरे ॥

Rig Veda 8.12.28

All planets remain stable because as they come closer to sun due to attraction, their speed of coming closer increases proportionately.

होळ चिदारुजत्नुमिर्गिहा चिदिन्दर वहिनम: ।
अविन्द उसिया अनु ॥

Rig Veda 8.16.5

यदा सूर्यममुं दिवि शुक्रं जयोतिरधारय: । आदित्ते व ...

Rig Veda 8.12.30

O God, You have created this Sun you possess infinite power.

You are upholding the sun and other spheres and render them steadfast by your power of attraction.

हेरण्यपाणि: सविता विचर्षणिर उभे दयावाप्रिथ्वी अन्तर ईयते ।
अपामीवाम वाधते वेति सूर्यम अभि कर्णेन रजसा दयाम रणोति ॥

Rig Veda 1.35.9

The sun moves in its own orbit but holding earth and other heavenly bodies in a manner that they do not collide with each other through force of attraction.

पञ्चारे चक्रे परिवर्तमाने तस्मिन्ना तस्थुर्भुवनानि विश्वा ।
तस्य नाक्षस्तप्यते भूरिभारः सनादेव न शीर्यते सनाभिः ॥

Rig Veda 1.163.13

Sun moves in its orbit which is moving. Earth and others move around sun due to force of attraction, because sun is heavier than them.

GRAVITATIONAL FORCE

All planets remain stable because as they come closer to sun due to attraction, their speed of coming closer increases proportionately. (Rig Veda 8.12.28)

O God, You have created this Sun. You possess infinite power. You are upholding the sun and other spheres and render them steadfast by your power of attraction. (Yajur Veda 33.43)

The sun moves in its own orbit in space taking along with itself the mortal bodies like earth through force of attraction. (Rig Veda 1.35.9)

The sun moves in its own orbit but holding earth and other heavenly bodies in a manner that they do not collide with each other through force of attraction. (Rig Veda 1.164.13)

Sun moves in its orbit which itself is moving. Earth and other bodies move around sun due to force of attraction, because sun is heavier than them. (Atharva Veda 4.11.1)

LIGHT OF MOON

The moving moon always receives a ray of light from sun. (Rig Veda 1.84.15)

Moon decided to marry. Day and Night attended its weddin Sun gifted his daughter "Sun ray" to Moon. (Rig Veda 10.85.9)

ECLIPSE

O Sun! When you are blocked by the one whom you gifted your own light (moon), then earth gets scared by sudden darkness.

Shloka:

हिरण्यपाणिः सविता विचर्षणिरुभे द्यावापृथिवी अन्तर ईयते ।
अपामीवाम् बाधते वेति सूर्यम् अभि कर्णेन रजसा द्यामृणोति ॥

Rig Veda 5.40.5

Translation:

हिरण्यपाणिः…the golden-handed; सविता…sun; विचर्षणिः…very active or busy;

उभे…fills; ईयते…so large; अन्तर…distance; द्यावापृथिवी…heaven and earth

सूर्यम्…sun; बाधते…drive away; अपामीवाम्…disease; वेति…sets in motion;

रणोतिअभि…penetrates through; कर्णेन रजसा…removes dark; द्याम्…via compassion.

The golden-handed Sun, the active, fills (with his rays) the large distance between the earth and heaven. He (the sun) drives away sickness, sets things in motion, penetrates through and removes darkness via his compassion.

There is definitely a hint to gravity here.

आदित्यो ह वै बाह्यः प्राण उदयत्येष ह्येनं चावशुषं प्राणमनुगृह्णानः ।
पृथिव्यां या देवता सैषा पुरुषस्य अपानमवष्टभ्यान्तरा यदाकाशः स समानो वायुर्व्यानः ॥८॥

Adityo ha vai bahyah prana udayatyesa hyenam caksusam pranamanugrhnanah |

Prthivyam ya devata saisa purusasyaapanamavastabhyantara

yadakasah sa samano vayurvyanah || 8 ||

8. The sun, indeed, is the external prana. He rises favouring the prana in the eye. So the goddess of the earth attracts the apana downwards. The akasa between is samana. The wind is vyana.

Why does an object held fall when you let go?

To describe this first of all Vaishesika Sutra 5.1.6 states:

आत्मकर्म हस्तसंयोगाक्ष ।

Action of body and it's members is also from conjunction with the hand.

As the above Sutra describes that it is due to conjunction with hand object remains. Then the next Sutra describes that in the absence of conjunction falling results due to Gravity.

संयोगभावे गुरुत्वात्पतनम्

Vaishesika Sutra (VS) (5.1.7)

In the absence of conjunction falling results from Gravity.

Thus it clearly recognises objects fall downward due to Gravity.

1) Why does an object thrown in air fall after sometime?

Then Vaishesika Sutra discuss role of Gravity in falling of moving objects. It gives through the analogy of arrow. First it gives mechanism of arrow projection in VS 5.1.17

नोदनाध्यभिषोः कर्म तत्कर्मकारिताच्च संस्काराद्उत्तरं तथोत्तरमुत्तरं च ॥

VS 5.1.17

The first action of arrow is from impulse; the next is resultant energy produced by the first action, and similarly the next next.

I ancient india important events such birth of Lord Rama, Lord Krishna etc,. were recorded as per position stats, lunar phsse and constellation.

The earliest calendar on record and still followed in Kashmir is Saptarishi Samvat. As per Gregarian Calender 2010 corresponds to 5786 of Saptarishi Samvat.

'Jin nivaanato pacchha pure tassaabi sekata Saashta rasan shatadvyamavan vyaaneeyam' Which means Samrat Ashoka was crowned in 6208.

Surya Siddhanta: Surya Siddhanta is first among traditions or doctrines (Siddhanta) in archeo-astronomy of vedic era. Infact it is the oldest ever book in the world which describes earth as sphere but not flat, gravity being reason for objects falling on earth etc. Going by calculations of Yugas (which will be dealt in detail in Manvantara theory) the first version of Surya Siddhanta must have been known around two million years old. However the present version of Surya Siddhanta that is available is believed to be more than 2500 years old, which still makes it the oldest book on earth in astronomy.

Few excerpts from Surya Siddhanta:

- The average length of tropical year as 365.2421756 days which is only 1.4 seconds shorter than modern value of 365.2421904 days!
- The average length of sidereal rotation (the rotation of earth referencing the fixed star) as 23 hours 56 minutes and 41 seconds, the modern value is 23 hours 56 minutes 4.091 seconds.
- Similarly value of length of sidereal year is 365 days 6 hours 12 minutes and 30 seconds which has an error of 3 minutes and 20 seconds.
- The Surya Siddhanta also states the motion and diameter of planets! For instance the estimate of diameter of Mercury is 3,008 miles, an error of lessthan 1% from current accepted diameter. It also estimates the diameter of Saturn as 73,882 miles which again has an error of less than 1%.
- Asidefrom inventing the decimal system, zero and standard notation (giving the ancient Indians the ability to calculate trillions when rest of the world was struggling with thousands. The Surya Siddhanta also contains the roots of Trignometry.
- It uses sine (jya), cosine (kojya or "perpendicular sine") and inverse sine (otkram jya) for the first time.

- Objects fall on earth due to force of attraction, therefore the earth, the planets, constellations; the moon and sun are held in orbit due to this attraction. This was also discussed in Prashno Upanishad (1.1 & 1.11).
- Surya Siddhanta also goes into a detailed discussion about time cycle and that time flows differently in different circumstances roots of relativity.
- The astronomical time cycle contained in the text was remarkably accurate at that time.
- Hindu Calender is called Panchanga. It is based on Surya Siddhanta. **Based** on Panchanga date of Birth and date of Death of Lord Krishna can be calculated. This shows antiquity of Surya Siddhanta.

Time of Lord Krishna's Birth

Krishna Birth – 18th July 3228 BCE

Krishna Death – 18th February 3102 BCE (the start of Kali Yuga)

According to above time lines Lord Krishna lived for 126 years and 5 months. If we had Gregorian calendar at the time of Lord Krishna's birth it would had been 23rd June –3227 on Gregorian calendar. DrikPanchang calculations for 23rd June –3227 and 24th June –3227 gives following Panchang data for Mathura location.

Panchang for 18th July 3228 BCE	Panchang for 17th July 3228 BCE
Sunrise = 05:40:32	Sunrise = 05:40:52
Sunset = 19:28:10	Sunset = 19:28:30
Moonrise = 24:50:01+	Moonrise = 25:18:05+
Moonset = 12:04:15	Moonset = 12:56:13
Month = **Shravana** (Amanta)	Month = **Shravana** (Amanta)
Tithi = **Ashtami upto 16:16:22**	Tithi = **Saptami upto 14:06:01**
Paksha = **Krishna Paksha**	Paksha = **Krishna Paksha**
Nakshatra = **Rohini upto 13:23:01**	Nakshatra = Krittika upto 10:38:15

Astronomy

Panchang for 18th July 3228 BCE	Panchang for 17th July 3228 BCE
Yoga = Vajra – 28:04:07+	Yoga = Harshana – 27:31:39+
Karna = Kaulava – 16s:16:22	Karna = Bava – 14:06:01
Karna = Taitila – 29:12:14+	Karna = Balava – 27:13:46+
Sunsign = Simha	Sunsign = Simha
Moonsign = Vrishabha – 26:36:58+	Moonsign = Vrishabha

Panchang data shows that on 18th July 3228 BCE during Nishita or Hindu midnight both **Ashtami Tithi** and **Rohini Nakshatra** were prevailing. It should be noted that team of DrikPanchang found 18th July 3228 BCE as birthdate of **Lord Krishna** from majority of sources from Internet and our team just ran DrikPanchang software to get Panchanga data for this historical day. We don't have any proof to corroborate 18th July 3228 BCE as birthday of Lord Krishna but we are sure that 18th July 3228 BCE was *Krishna Janmashtami* day.

Ancient Greeks: They were the driving force behind the development of western astronomy and science. In many older textbooks, the Ancient Greeks are often referred to as the fathers of ancient astronomy, developing elegant theories and mathematical formulae to describe the wonders of cosmos.

The knowledge of Greeks was built upon solid foundations laid by other great cultures such as The Mesopotamian and Zorastrian astronomers. Greeks also lay at the crossroads of many trade routes, so ancient knowledge from the Indian Vedas and Chinese astromoners further contributed to the store of insight accumulated by the Greeks.

Sumerians: Whilst we can safely assume that humanity developed Sophisticated astromonical techniques long before the dawn of recorded history, the history of western astronomy begins with Mesopotamia. This land, straddling the fertile cresent between the Tigris, Euphrates rivers, which now lie in Iraq, Turkey, Syria and Iran. The Fertine Cresent is where civilization began, and was home to the great civilization of the Sumerians, Babylonians and Assyrians.

Historians are sure that the oldest of the civilizations, the Sumerians, were astronomers, but most of their knowledge was lost, leaving a few tantalizing fragments of their sophisticated culture. Our knowledge of their contribution to ancient astronomy is gleaned from indirect transmission to the Babylonian culture. Early Babylonian records, dating around 1800 BC, use the Sumerian names of stars, suggesting that the body of knowledge was passed down.

Babylonian Ancient Astronomy: The first real insight into the history of astronomy begins with the Babylonians, who used the heavens as their means to establish an accurate calender which was useful for agriculture. From 1800 BC they accurately plotted the movement of the Sun and the Moon, using them to track the procession of seasons.

Many notable events in history were shaped by conjunction of stars and astronomic events, such as the star (supernova) that guided the magi to Bethlehem. Bible teaches you the way to heaven, not the way heavens move. Halley's Comet always mistaken to bring fear and destruction.

Modern Astronomy timeline	
130 BC –AD 1400 Stellar Geometry	Early star catalogues
Before 1600 Solar	Guesses of the distance of the Sun
Late 1500s Solar System Geometry	Brahe
c.1610 Astrophysics	Kepler discovers the rules of planetry motion
1660 Astrophysics	Newton's theories of dynamics and gravity
1676 Solar System Geometry	Romer measures the speed of light
1776 Astrophysics	Maskelyne weighs the Earth
1810 Stellar Geometry	Bessel discovers the parallex of a distant star
19th century Solar System Geometry	Transits of Venus
1901 Solar System Geometry	Eros geometrical measure
1905–1913 Cosmology	Einstein's theories of relativity

Modern Astronomy timeline	
1926 Astrophysics	Eddington makes the sun shine
1930 Cosmology	Movement and structure of galaxies.
1990	Hubble Space Telescope
Satellites	
Space travel	

CHAPTER 3

PRAMANA

(Journal of Physics. Published by Springer Science & Business Media on Behalf of the Indian Academy of Sciences and in Collaberation with The Indian National Science Academy and Indian Physics Association)

INTRODUCTION

We live in the age of evidence; be it judiciary, scientific inquiry/research or in medical practise. Pramana literally means 'proof' or 'evidence.' It refers to epistemology (means knowledge or logas). Epistemology is a branch of philosophy that investigates into the origin, nature, methods and limits of the human knowledge. Epistemology is one of the key, much debated fields of study in Sanatana Dharma, Buddhism and Jainism since ancient times.

SANATANA DHARMA

Sanatana dharma means Eternal Law (Dharma equivalent in Latin is– firmus firm.) Sanatana Dharma identifies six pramanas as correct means of accurate knowledge. Pratyaksha (perception), Anumana (inference), Upamana (comparison and analogy), Arthapatti (postulation, derivation from circumstances), Anupalabdi (nonperception, negation, cognitive proof) and Sabda (word, testimony of past or present by reliable experts). In some other text such as by Vedavyasa ten pramanas are discussed.

i. ***Pratyaksha***[1] means perception. It is of two types: external and internal. External perception is described as that arising from interaction of five senses and worldly objects, while internal perception is described as that of internal self, the mind. The ancient texts identify four requirements for correct perception.

   a. Indriyarthasannikarsa: this is direct experience by one's sensory organ(s) with the object, of whatever being studied.

   b. Avyapadasya: non-verbal, correct perception is not through here-say, according to ancient Indian scholars, where one's sensory organs rely on accepting or rejecting someone else's perception.

   c. Avyabhicara: does not wander correct perception, does not change, nor is a result of the deception because one's sensory organ. It excludes if one's observation is drifting, defective, or suspect.

   d. Vyavasayatmaka: (definite dot and comma, correct) perception excludes judgement of doubt, either because of one's failure to observe all the details, or because one is mixing inference with observation and observing what one wants to observe, or not observing what one does not want to observe. Some ancient scholars propose "unusual perception" as pramana and called it internal perception, a proposal contested by other Indian scholars. The internal perception concept included Pratibha (intuition),

   Samanyalaksana-pratyaksa (a form of induction from perceived specifics to universal), and jnanalaksanapratyaksa (a form of perception of prior process and previous states of a topic of study by observing its current state. Further some

---

[1] I.Kumar, Anil, Prakasam, E.R. Kalyane Vijay (2008)

schools of Sanatana Dharma considered and refined rules of accepting uncertain knowledge of Pratyaksha – Pramana, so as to contrast nirnaya (definitive judgement, conclusion) from Anadhyavasaya (indefinite judgement).

ii. ***Anumana*** means inference. It is described as reaching a new conclusion and truth from one or more observations and previous truths by applying reason. Observing smoke and inferring fire is an example on Anumana. In all except one Sanatana Dharma philosophies, this is a valid and useful means of knowledge. The method of inference is explained by Indian texts as consisting of three parts: Pratigna (hypothesis), Hetu (reason) and drshtana (examples). The hypothesis must further be broken down into two parts, state the Indian scholars: Sadhya (that idea which needs to be proven or disproven) and Paksha (the object on which the Sadhya is predicted). The inference is conditionally true if Sapaksha (positive examples as evidence) are present, and if Vipaksha (negative examples as counter-evidence) are absent. For rigor, the Indian philosophy also state further epistemic steps. For example, they demand Vyapti-the requirement that the Hetu (reason) must necessarily separately account for inferences in "all" cases, in both Sapaksha and Vipaksha.

A conditionally proven hypothesis is called Nigamana (conclusion).

iii. ***Upamāna*** means comparison and analogy. Some Sanatana schools consider it as a proper means of knowledge. Upamāna may be explained with the example of a traveller who has never visited the lands or islands with endemic population of wildlife. He or she is told, by someone who has been there that in those lands you see an animal that sort of looks like a cow, grazes like a cow but is different from the cow in such and such a way. Such use of analogy and comparison is; state that Indian epistemologists, a valid means of conditional knowledge, as it helps the traveller identify the new animal later. The subject of comparison is formally called Upameyam, the object of comparison is called Upamanam, while the attribute(s) are identified as Samanya.

Thus explains Monier Williams, if a boy says "her face is like the moon in charmingness", her face is the Upameyam, the moon is Upamanam and charmingness is Samanya. The 7th century text Bhattikavya in verse 10.28 through 10.63 discusses many types of comparisons and analogies, identifying when this epistemic method is more useful and reliable and when it is not in the various ancient and medieval texts of Hinduism, 32 types of Upamana and their value in epistemology are debated. The detailed description of the above is beyond the scope of the present presentation. Interested readers can get full knowledge by reading further literature to the effect.

iv. **Arthapatti** means postulation, derivation from circumstances. In contemporary logic, this Pramana is similar to circumstantial implication. As example, if a person left on a boat in the river earlier, and the time is now past the expected time of arrival, then the circumstances support the truth postulate that the person has arrived. Many Indian scholars considered this Pramana an invalid or at the best weak, because the boat may have gotten delayed or diverted. However, in cases such as deriving the time of future sunrise or sunset, this method was asserted by proponents to be reliable.

Another common example for Arthapatti in ancient Sanatana texts is

That if "Devadatta is fat" and "Devadatta does not eat in day" then the following must be true; "Devadatta eats in the night." This forms postulation and deriving from circumstances is, claim the Indian scholars, a means to discovery, proper insight and knowledge. The Hindu schools that accept this means of knowledge state that this method is a valid means to conditional knowledge and truths about the subject and object in original premises or different premises. The schools that do not accept this method, state that postulation, extrapolation and circumstantial implication is either derivable from other Pramanas or flawed means to correct knowledge, instead one must rely on direct perception or proper inference.

v. *Anupalabdi* means non-perception, negation/cognitive proof. Anupalabdi Pramana suggests that knowing a negative, such as "there is no jug in the room" is a form of valid knowledge. If something can be observed or inferred or proven as non-existent or impossible, then one knows more than what one did without such means. In the two schools of Sanatana Dharma that considers Anupalabdi as epistemically valuable, a valid conclusion is either Sadrupa (positive) or Asadrupa (negative) relation both correct and valuable. Like other Pramana Indian scholars refined Anupalabdi to four types: non-perception of the cause,

Non perception of the effect, non-perception of the object and non-perception of contradiction. Only two schools of Sanatana Dharma accepted and developed the concept "non-perception" as Pramana. The school that endorsed Anupalabdi affirmed that it is a valid and useful when other five Pramanas fail in one's pursuit of knowledge and truth. Abhava means non-existence. Some scholars consider Anupalabdi to be same as Abhava, while others consider Anupalabdi and Abhava as different. Abhava Pramana has been discussed in ancient Hindu context of Padartha. A Padartha is defined as that which is simultaneously Astitva (existent). Jneyatva (knowable) and Abhidheyatva (namable). Specific examples of Padartha, include Dravya (substance), Guna (quality), Karma (activity), Samanya/Jati (universal/class property), Samavaya (inherence) and Vishesha (Individuality). Abhava is then explained as "referents of negative expression" in contrast to "referents of positive expression" in Padartha. An absence, state the ancient scholars, is also "existent, knowable and namable", giving the example of negative numbers, silence as a form of testimony, Asatkaryavada theory of causation, and analysis of deficit as real and valuable. Abhava was further refined in three types that accepted it as a useful method of epistemology: Dhvamsa (termination of what existed), Atyanta-Abhava (impossibility), Pragavasa (prior, antecedent non-existence).

vi. **Sabda** (word) means relying on word, testimony of past or present reliable experts. Hiriyanna explains Sabda-Pramana as a concept which means reliable expert testimony. The Santana schools which consider it epistemically valid suggest that a human being needs to know numerous facts, and with the limited time and energy available he can learn only a fraction of those facts and truths directly. He must rely on others, his parents, family, friends, teachers, ancestors and kindred members of society to rapidly acquire and share knowledge and thereby enrich each other's lives. This means gaining proper knowledge is either spoken or written, but through Sabda (words). The reliability of source is important and legitimate knowledge can only come from Sabda (word) or reliable source. Now the problem comes as to how to establish reliability.

CARVAKA SCHOOL

Carvaka school accepted only one valid source of knowledge ie., perception. Itheld all remaining methods as outright invalid or prone to error and therefore invalid. Carvaka is originally known as Lokayata and Brhaspatya is an ancient school of Indian materialism' Carvaka holds direct perception, empiricism and conditional inference as proper sources of knowledge, embraces philosophical scepticism and rejects Vedas, Vedic ritualism and supernaturalism.

Brihaspati is usually referred as the founder of Charvaka or Lokayata Philosophy. Much of the primary literature of Barhaspatya Sutra (ca 600 BCE) are missing or lost. Its teachings have been compiled from historic secondary literature such as those found in the Shastras, Sutras and the Indian epic poetry as well as in dialogues of Gautama Buddha and from Jain literature.

Charvaka out rightly rejects inference as means to establish valid universal knowledge and metaphysical truths. In other words Charvaka epistemology states that whenever one infers a truth from a set of observations or truths, one must acknowledge doubt that inferred knowledge is conditional.

Charvaka is categorized as heterodox school of Indian philosophy. It is considered as example of atheistic schools in the Hindu tradition. As it has been mentioned previously Sanata Dharma encourages to question and also has place for atheists. It embraces all knowledge and has a rich tradition. It teaches you to be fearless.

Charvaka school of Hinduism did not believe in karma, rebirth or an afterlife. The Sarvasidhanta Samgraha states the Charvaka position as follows:

> **There is no other world other than this**
> **There is no heaven and no hell**
> **The realm of Shiva and other regions**
> **are invented by stupid imposters.**
>
> — **Sarvasidhanta Samgraha, verse 8**

Charvaka believes that there is nothing wrong with sensual pleasure. Charvaka thought that wisdom lay in enjoying pleasure and avoiding pain as far as possible. Charvaka did not believe in austerities or rejecting pleasure out of fear of pain and held such reasoning as foolish.

The Sarvasidhanta Samgraha states that Chavarka position on pleasure and hedonismas follows:

"The enjoyment of heaven lies in eating delicious food, keeping company of young women, using clothes and perfumes, garlands, sandal paste – while Moksha is death which is cessation of life breath – the wise therefore ought not to take pains on account of Moksha.

A fool wears himself out by penances and fasts. Chastity and other such ordinances are laid down by clever weaklings.

– Sarvasidhanta Samgraha verse 9-12

Charvakas according to Sarvadarsanasamgraha verses 10-11, declared the Vedas to incoherent rhapsodies whose only usefulness was to provide livelihood to priests. They also held the belief that Vedas were invented by man and had no divine authority.

VAISHESHIKA SCHOOL

Introduction: Vaisheshika philosophy is closely associated with the Nyaya philosophy, which together form one of the six orthodox Vedic systems from ancient times. The six sidhantas are viz,: i) Nyaya, ii) Vaisesika, iii) Samkhya iv) Yoga, v) Purva mimamsa and vi) Uttara mimamsa or Vedanta. It adopts a form of atomism and contends that every object in the physical universe is reducible to a finite number of atoms. This school of thought was expounded by Sage Kanada in 6th century BC. Nyaya and Vaisheshika are considered sister philosophies. It emphasises that a virtuous life guided by principals of Dharma enables an individual to achieve a fulfilling life which has the spirit of highest good or liberation. The two systems of philosophy maintain a dualistic concept which encompasses God and Jiva.

Vaisheshika believed that only perception and inference are of prime importance. This is a theistic form of atomism. Vaisheshika system developed a theory to explain the properties of material as the interaction of different types of atoms that make up the material.

In Sankhya, Yoga, Vishistadvaita Vedanta and Dvaita Vedanta schools, the proper means of knowledge must rely on these three Pramanas:

1. Pratyaksa – Perception
2. Anumana – Inference
3. Sabda – Testimony/word of reliable experts.

Nyaya School accepts four means of obtaining knowledge (Pramana), e.i., Perception, Inference, Comparison and Word.

Purva Mimamsa school

Mimanansa means "reflection" or "critical" investigation. It is also one of the six orthodox schools of Vedanta. It has deeply influenced the formation of Hindu Law. The aim of Mimamsa is to give rules for interpretation of Vedas.

In Mimansa school of Sanatana Dharma the following Pramansas are considered proper:

1. Pratyaksha
2. Anumana
3. Sabda (word)
4. Upamana
5. Arthapatti (postulation, presumption)

Advaita Vedanta and Bhetta Mimamsa schools: In this the following Pramanas are accepted:

1. Pratyaksha
2. Anumana
3. Sabda (word)
4. Upamana
5. Arthapatti
6. Anupalabdi, Abava (non perception, cognitive proof using non-existence)

Buddhism

Strictly speaking, Pramana (tshadma) means valid cognition. In Buddhism it refers to the tradition, principally associated with Dignaga and Dharmakirti of logic (rtags rigs) and epistemology (blo rigs).

Buddhism accepts only two Pramana (tshad ma) a valid means to knowledge:

Pratyaksha and Anumana: Buddhism also considers scriptures as third valid Pramana as from Buddha and other "valid minds" and "valid persons."

Purport: The above expositions by various schools of thought are to arrive at Truth through various means. The purpose of the above writing helps us in understanding creation of Universe based on the Pramana by different schools which will be dealt exhaustively in subsequent chapters.

Nyaya

1. Perception called Pratyaksa occupies the foremost position in Nyaya epistemology. Perception defined by sense – object contact and is unerring. Perception can be of two types – ordinary or extraordinary.

    Ordinary (Laukika or Sadhana) perception which is of six types, i.e.,
    - i) Visual by eyes
    - ii) Olfactory by nose
    - iii) Auditory by ears
    - iv) Tactile by skin
    - v) Gustatory by tongue
    - vi) Mental by mind

Extraordinary (Alaukika or Asadharana) perception is of three types i.e.,

   i) Samanyalaksana (perceiving generality from particular object).

   ii) Jnanalaksana (when one sense organ can also perceive qualities not attributable to it, as when seeing s chilli, one Knows that it would be hot).

   iii) Yogana (when certain human beings, form the power of yoga, can perceive past, present and future and have supernatural abilities, either complete or some).

2. Inference: called Anumana is one of the most important contributions of Nyaya. It can be of two types – inference for oneself (Svanthanumana, where one does not need any formal procedure and at the most the last three of their five steps).

3. Comparison, called Upamana. It is produced by the knowledge of resemblance or similarity, given some pre-description of the new object beforehand.

4. Word or Sabda are also accepted as Pramana. It can be of two types:

   i. Vaidika (Vedic)

   ii. Laukika or words

CHAPTER 4

SĀMKHYA/SĀNKHYA SUTRA KAPILA

(Translated by James R. Ballantyne, LL.D)

INTRODUCTION

Sānkhya is derived from the word 'Sāmkhya' meaning a sense of thinking and counting. Of all the philosophical systems Sānkhya philosophy is considered to be the most ancient school of thought. Sānkhya philosophy maintains a prominent place in all the Sastras since it is either controverted or supported by every other philosophical system. Sānkaracarya says "This doctrine, moreover, can stand somewhat near to Vedanta doctrine since like the latter; it has been accepted by some of the authors of the Dharma Sutras, such as Devala and so on. For all these reasons Kapila who authored Sānkhya Sutra has taken special trouble to refute the Pradhana Doctrine."

Sāmkhya forms one of the most important pillars consisting the six systems (Siddhantas) of Indian philosophy. Its contribution to our knowledge of Reality and the world is seminal. Today Vedanta rules the roost and modern science is finding itself more and more in agreement with the intuitive perceptions of this sixth darsana i.e.: Samkhya Sutra but it must be noted that Vedanta takes off to the ethereal heights only from the granite platform provided by the standard works Samkhya

Sutra. As of today only Samkhya Sutra, SamkhyaTattva & Sankhya karika available.

Vedanta accepts most of the basic concepts of Samkhya; the three gunas of Prakriti in terms of which can be explained not only the manifold objects of the universe, but also workings of the mind and psyche and even the rationale of medical therapy; the process of evolution, long before Western science began to think in terms of it. Samkhya reduces everything to two entities – Prakriti (nature) and Purusa (God). Coming to three Gunas viz Sathvik (goodness), Rajas (Passion) and Tamas (Ignorence).[2] Lot has been said by Lord Krishna in 14th chapter of Srimad Bhagwat Gita. It is devoted to 3 gunas but with spiritual message.

सांख्ययोगौ पृथग् बालाः प्रवदन्ति न पण्डिताः ।
एकमप्यास्थितः सम्यगुभयोर्विन्दते फलम् ॥

(Ch 5 Verse 4 Srimad Bhagwat Gita)

Only the ignorance speaks of devotional service (Karma – yoga) as being different from the analytical study of the material world (Sankhya). Those who are actually learned say that he who applies himself well with one of these paths achieves the result of both.

सत्त्वं रजस्तम इति गुणाः प्रकृतिसम्भवाः ।
निबध्नन्ति मसाबाहो देहे देहिनमव्ययम् ॥

(Ch 14 verse 5 Srimad Bhagwat Gita)

Sattvam – the mode of goodness; rajah – the mode of passion; tamas – the mode of ignorance; iti-thus; gunah – the qualities; Prakriti – the material nature; sambhavah – produced of; nibadhnanti – do condition; maha – baho – O mighty armed one; dehe – in this body; deham – the living entity, avyayam – eternal.

---

[2] Gerald James (2011), Classical Samkhya: Interpretation of its History and Meaning, Motilal Banarsidas ISBN 978-206

Material nature consists of three modes – Sathvik, Rajas and Tamas viz goodness, passion and ignorance. When the eternal entity comes into contact with nature (Prakriti), O mighty armed Arjuna, he becomes conditioned by the three modes.

Sage Kapila is traditionally credited as founder of the Samkhya School. The exact period in which Samkhya was written is unclear. It may predate Rig Veda for it gets mentioned in Rigveda eg Tamas is described in Rigveda as:

तमआसीत् तमसा गूळ्हमग्रे

(RV 10.129.3)

Here 'tamas' has been described as darkness.

Which later assumed the form of unmanifest? This very Rigveda shows the dissolution of elemental world in its cause, thus indicating Satkara Vada to which philosophy Sankhya belongs. Even the Pradhana is referred to as Aja (unborn) and Veda explains it as below: (RV 10.82.6)

तमिद्गर्भं प्रथमं दध्र आपो
यत्र देवाः समगच्छन्त विश्वे
अजस्य नाभावध्येकमर्पितं
यस्मिन् विश्वानि भुवनानि तस्थुः ॥

(RV 10.82.6)

Waters first bore this child of womb where all the Gods contemplated. The One (Ekam) is established (arpitam) in the centre (nabhid) of the unborn (ajasya) [waters] and there is that One all the worlds abide. "This truth is one (ekam); it is the centre of the Primieval water which has no birth (aja) i.e., One itself created Water and stayed in it.

Further, the Sattva, Rajas and Tamas of Sankhya Philosophy are explained in Chandogya Upanishad and in Katha Upanishad (3.10.11). It is a well known fact that Svetasvatara Upanishad essentially a Sankhya Upanishad because it clearly mentions the Sankhya categories.

Lord Krishna while describing some of his splenderous manifestation thus says in Srimad Bhagwat Gita:

सिध्दानां कपिलो मुनि:

(Bh G 10.26)

Lord Krishna said 'of all those who are perfected I am the sage Kapila Muni.' This is to show the great importance being shown to sage Kapila.

Samkhya is most related to Yoga Sidhanta. Samkhya is an enumerationist philosophy whose epistemology accepts three of six pramanas as only reliable means of gaining knowledge. These include Pratyaksha (Perception), Anumana (Inference) and Sabda (aptavacana, word/testimony of reliable sources).[3, 4, 5] Sometimes described as one of the rationalist school of Indian philosophy, this ancient school Samkhya strongly is dualistic.[6,7,8] Samkhya philosophy regards the universe as consistent of two realities; Purusa (Consciousness) and Prakriti (matter). Jiva the living being is that state in which Purusa is bonded to Prakriti in some form. This fusion of Purusha and Prakriti, state the Samkhya Scholars, lead to emergenceof Buddhi (intellect) and Ahankara (Ego consciousness)

Samkhya and universe:

In Samkhya the Universe described by this school as one created by Purusa – Prakriti entities infused with various permutations and combinations of various enumerated elements, senses, feelings, activity

---

[3] Larsen 1998, p. 9

[4] Eliott Deutsche (2000), in Philosophy of Religion: Indian Philosophy Vol 4 (Editor: Roy perrot), Routledge, ISBN 978-0815336112, pages 245-248. John A. Grims, A concise Dictionary of Indian Philosophy: Sanskrit terms Defined in English, State, University of New York Press, ISBN 978-079140675, and page 238.

[5] John A. Grimes, Concise Dictionary of Indian Philosophy: Sanskrit Term Defined in English, State University of New York Press, ISBN 978-7091430675, page 238.

[6] Michel 2014, P. 264

[7] Sen Gupta 1986 p6

[8] Radhakrishna and more 1957, p.89

and mind. During the state of imbalance, one or more constituents overwhelm the others, creating a form of bondage, particularly of the mind. The end of this imbalance, bondage is called Liberation or Kaivalya by the Samkhya School. The existence of God or Supreme Being is not directly asserted, nor considered relevant by Samkhya philosophies. Sankhya denies the final cause as Iswara (God).[9]

While the Samkhya School considered the Vedas as reliable source of knowledge, it is an atheistic philosophy according to Paul Deussen and other scholars.[10,11] A key difference between Samkhya and yoga school, state scholars is that yoga school accepts "personal, yet essential inactive, deity" or "personal God."[12]

The Samkhya system is based on Sat-Karya-vada or theory of Causation. According to Sat-Karya-vada, effect is pre-existent in the cause. There is only an apparent or elusory change in make up of the cause and not material one, when it becomes effect.

असदकरणदुपादानग्रहणात् सर्वसंभवाभावत् ।
शक्तस्य शक्यकरणानत् कारणभावच्च सत् कार्यम् ॥९॥

Aphorism 9 from Samkya Karika.

The effect is ever existent, because

1. What is non-existent can by no means be brought into existence
2. Because effects take adequate material cause;
3. Because all effects are non producible from all causes
4. Because an efficient cause can produce only that for which it is efficient and finally
5. Because the effect is of same essence as the cause.

---

[9] Dasgupta 1922, p.258

[10] Mike Burley (2012), Classical Samkhya and Yoga-An Indian Metaphysics of Experience, Routledge, ISBN 978-0415648875, page 39.

[11] Loyd Pflueger, Person Purity and Power in Yoga Sutra, in Theory and Practice of Yoga (Editor Knut Jacobson), Motilal Banarasidas, ISBN 978-8120832329, Page 38-39

[12] Kovoor T. Behanan (2002), Yoga its Scientific Basis, Dover, ISBN 9780486417929, page56-58

The following are the proofs that establish non differences of the effect from the cause.

1. The cloth is not different from the yarn (constituting it) because cloth subsists in the yarn. The differing in its essence from another, cannot subsist in it, like a cow in horse, but here, the cloth subsists in its yarn. From this it follows that the effect is not different from its cause.
2. The cloth and yarn cannot be two different things because of the relationship between the material cause and the effect (Apadena-upadeya bhava). Whenever two things are found to be different from each other, there is a relationship between the constituent cause and effect is never found, that is in the case of the jar and the cloth. But relationship between the constituent cause and the effect is found between the yarn and the cloth, the two are not different things.
3. For the following reason also cloth and yarn are not two defferent things because there is neither conjunction nor disjunction between them (Samayoga-aprapti-abhavat). Conjunction is found to exist between object different from each other, as between the well and jujube tree (a kind of tree), the same with regard to separation also, as between the Himavan mountain and Vindhya mountains(These mountains are in India).
4. For the following reason also, the cloth does not differ from the yarn because the cloth does not contain in itself any other product which make it heavier than the yarn.

Atheism and Samkhya:

Samkhya accepts the notion of higher selves or perfected beings but rejects the notion of God. Classical Sankhya argues against the existence of Ishwara[13] on the metaphysical grounds. Samkhya theorists argue that

---

[13] Jerald James Larson (2011), Classical Samkhya... An Interpretation of its History and Meaning, Motilal Banarasidas, ISBN 978-8120805033, Pages 154-206.

an unchanging God cannot be source of an everchanging world and that God was only a necessary metaphysical assumption demanded by circumstances. The Sutras of Samkhya have no explicit role for a separate God distinct from the Purusha. Such a distinct God is inconceivable and self-contradictory and some commentaries speak plainly on this subject.

Arguments against Ishvara' Existence

The following arguments given by the Samkhya philosophy is against the idea of an eternal, self cause, Creator God.

- If the existence of Karma (work – good or bad) is an assumed, the proposition of God as a moral Governor[Aph 3 book V] of universe is unnecessary. For, if God enforces the consequences of action then He cannot do so without Karma. If however, He is assumed to be within the law of Karma, then Karma itself would be giver of consequences and there would be no need of God.
- Even if Karma is denied, God still cannot be enforcer of consequences. Because the motives of an enforcer God would either egoistic or altruistic. Now, God's motive cannot be assumed to be altruistic because an altruistic God would not create world so full of suffering. If His motives are assumed to be egoistic, then God must be thought of to have desire, as agency or authority cannot be established in the absence of desire. However, assuming that God has desire would contradict God's eternal freedom which necessitates no compulsion of activity. Moreover, desire according to Samkhya is an attribute of Prakriti and cannot be thought to grow in God. The testimony of the Vedas according to Samkhya also confirms this notion.
- Despite arguments to contrary, if God is still assumed to contain unfulfilled desires, this would cause Him to suffer pain and other similar human experiences [Aph 4 Book 5] Such a worldly God would be no better than Samkhya's notion of higher self.

- Furthermore, there is no proof of existence of God. He is not the object of perception, there exists no general proposition that can prove Him by influence and the testimony of the Vedas speak of Prakriti as orogin of world, not God. Therefore, Samkhya maintained that the various cosmological and teleological arguments could not prove God.

वत्सविवृद्धिनिमित्तं क्षीरस्य यथा प्रवृत्तिरज्ञस्य ।
पुरुषविमोक्षनिमित्तं तथा प्रवृत्ति प्रधानस्या ॥५७॥

Samkhya Karika 57

Just as secretion of milk which is unintelligent, is for sake of the nourishment of the calf, similar is the action of Pradhana for sake of release of Purusa.

It is seen that insistent entities also act towards a definite purpose, eg milk, though inert, acts in the form of flowing for the nourishment of the calf. In a similar way Prakriti, though in sentient to the Purusa. It cannot however, be maintained that secretion of milk takes place under the superintevidence of God. Now, this action cannot be an instance vitiating the general proposition that action of inert things are due to the super-intendence of sentient beings; because the activity of every intelligent being always proceeds from either selfishness or compassion. In regard to the creation of universe, neither of this could be applied. From this it is clear that this creation cannot be due to the action of a sentient agent.

The Lord who is all-full, having all desires fulfilled, wanting in nothing whatsoever, cannot have any selfish motive in creating this world. Nor can it be said that Lord created the world out of compassion, because, compassion implies desire to alleviate other's pains; but prior to creation the spirits would have no bodies, organs and objects and consequently, no pain, no suffering. Then to remove whose pain would the compassion of the Lord be roused? If it be said that the pain subsequent to creation is the cause of compassion, then it would lead to the vitiating position of interdependence that creation is due to compassion and compassion is due to creation! Further if the creation was an act of compassion on part of Isvara (God), one would wish, Isvara to create only happy mortals, and not beings with variegated experiences.

Some of the interesting Sankhya Sutra (Aphorisms) in support of his view:

Sankhya Sutra Aphorisms from Book I

न कालयोगतो व्यापिनो नित्यस्य सर्वसं-बन्धात् ॥१२॥

| Time, which applies to all, cannot be the cause of the bondage of a part. | Aph. 12 Not from connexion with *time* [does bondage befall the soul]; because this, all-pervading and eternal, is [eternally] associated with *all*, [and not with those alone who are in bondage]. |

a. The bondage of man is not caused by *time*; because [if that were the case,] there could be no such separation as that of the *liberated* and *unliberated*; because time, which applies to everything, and is eternal, is at all times associated with all men, [and must, therefore, bring *all* into bondage, if any].

न देशयोगतोऽप्यस्मात् ॥१३॥

| Place, for the same reason, cannot be the cause. | Aph. 13 Nor [does bondage arise] from connexion with *place*, either, for the same [reason]. |

a. That is to say: bondage does not arise from connexion with *place*. Why? 'For the same reason,' i.e., for that stated in the preceding aphorism, viz., that, since it [viz., place] is connected with *all* men, whether liberated or not liberated, bondage would [in *that* case] befall the *liberated*, also.

नावस्यातो देहधर्मत्वात्तस्याः ॥१४॥

| The soul is not kept in bondage by its being conditioned. | Aph. 14 Nor [does the bondage of the soul arise] from its being conditioned [by its standing among circumstances that clog it by limiting it]; because *that* is the fact in regard to [not the soul, but] the *body*. |

a. By 'condition' we mean the being in the shape of a sort of association. The bondage [of the soul] does not arise from *that*; because *that* is the property of the *body* [and not of the soul]; because, that is to say, bondage might befall even the liberated [which is impossible], if that which is the fact in regard to another could occasion the bondage of one quite different.

b. But then [some one might say], *let* this conditioned state belong to the soul. On this point [to prevent mistakes], he declares:

श्रसङ्गोऽयं पुरुष इति ॥१५॥

| The soul is absolute. | Aph. 15 Because this soul is [unassociated with any conditions or circumstances that could serve as its bonds, it is] absolute. |

a. The word *iti* here shows that it [i.e., the assertion conveyed in the aphorism] is a *reason*; the construction with the preceding aphorism being this, that, *since* the soul is unassociated, it belongs only to the body to be conditioned.

प्रकृतिनिबन्धनाच्चेन्न तस्या श्रपि पारतन्त्यम् ॥१८॥

| Nature is not the immediate cause of the soul's bondage. | Aph. 18 If [you say that the soul's bondage arises] from Nature, as its cause, [then I say] 'no;' [because] that, also, is a dependent thing. |

a. But then [some one may say], let bondage result from *Nature*, as its cause. If you say so, I say 'no;' because that, also, i.e., Nature, also, is dependent on the *conjunction* which is to be mentioned in the next aphorism; because, if it [Nature] were to occasion bondage, even *without* that [conjunction which is next to be mentioned], then bondage would occur even in such cases as the universal dissolution, [when soul is altogether disconnected from the phenomenal].

b. If the reading [in the aphorism] be *nibandhaná* [in the 1st case, and not in the 5th], then the construction will be as follows: 'If [you say that] the bondage is caused by Nature,' &c.

c. Therefore, since Nature can be the cause of bondage, only as depending on something else [i.e., on the conjunction to be mentioned in the next aphorism], through this very sort of conjunction [it follows that] the bondage is *reflexional*, like the heat of water due to the conjunction of fire; [water being held to be essentially cold, and to *seem* hot only while the heat continues in conjunction with it].

d. He establishes his own tenet, while engaged on this point, in the very middle [of his criticisms on erroneous notions in regard to the matter; for there are more to come]:

न नित्यशुद्धबुद्धमुक्तस्वभावस्य तद्योगस्तद्योगादृते ॥१९॥

| What really is the relation of its bondage to the soul. | Aph. 19 [But] not without the conjunction thereof [i.e., of Nature] is there the connexion of that [i.e., of pain] with that [viz., the soul,] which is ever essentially a pure and free intelligence. |

a. Therefore, without the conjunction thereof, i.e., without the conjunction of Nature, there is not, to the soul, any connexion with that, i.e., any connexion with bondage; but, moreover, just through that [connexion with Nature] does bondage take place.

b. In order to suggest the fact that the bondage [of the soul] is reflexional [and not inherent in it, either essentially or adventitiously], he makes use of the indirect expression with a double negative, ['not without']. For, if bondage were produced by the conjunction [of the soul] with Nature, as colour is produced by heating [in the case of a jar of black clay, which becomes red in the baking], then, just like that, it would continue even after disjunction therefrom; [as the red colour remains in the jar, after the fire of the brick-kiln has been extinguished, whereas the red colour occasioned in a crystal vase by a China-rose, while it occurs *not without* the China-rose, ceases, on the removal thereof]. Hence, as bondage ceases, on the disjunction [of the soul] from Nature, the bondage is merely reflexional, and neither essential [§ **5. b.**] nor adventitious [§ **11. b.**].

c. In order that there may not be such an error as thot of the Vaiśeshikas, viz., [the opinion that there is] an absolutely real conjunction [of the soul] with pain, he says 'which is ever,' &c. [§ *19*]. That is to say: as the connexion of *colour* with essentially pure crystal does not take place without the conjunction of the China-rose [the hue of which, seen athwart the crystal, seems to belong to the crystal], just so the connexion of *pain* with the soul, ever essentially pure, &c., could not take place without the conjunction of some accidental associate; that is to say, pain, &c., cannot arise spontaneously, [any more than a red colour can arise *spontaneously* in the crystal which is essentially pure].

d. This has been declared, in the *Saura*, as follows 'As the pure crystal is regarded, by people, as red, in consequence of the proximity of something [as a China-rose] that lends its colour, in like manner the supreme soul [is regarded as being affected by pain].'

e. In that [aphorism, 19], the perpetual purity means the being ever devoid of merit and demerit; the perpetual intelligence means the consisting of uninterrupted thought; and the perpetual liberatedness means the being ever dissociated from *real* pain: that is to say, the connexion with pain in the shape of a *reflexion* is not a real bondage, [any more than the reflexion of the China-rose is a real stain in the crystal].

f. And so the maker of the aphorism means, that the cause of its bondage is just a particular *conjunction* [*§19. c.*]. And now enough as to that point.

g. Now he rejects [*§18. d.*] certain causes of [the soul's] bondage, preferred by others:

Some interesting Aph from Book I of Sam S

नाविद्यातोऽप्यवस्तुना बन्धायोगात् ॥२०॥

The Vedántic tenet on this point disputed.	*Aph.* 20 Not from Ignorance, too, [does the soul's bondage arise]; because that which is not a reality is not adapted to binding.

a. The word 'too' is used with reference to the previously mentioned 'Time,' &c., [*§ 12*, which had been rejected, as causes of the bondage, antecedently to the statement, in *§ 19*, of the received cause].

b. Neither, too, does [the soul's] union with bondage result directly from 'Ignorance,' as is the opinion of those who assert non-duality [or the existence of no reality save one (see *Vedánta-sára, § 20. b.*)]; because, since their 'Ignorance' is not a real thing, it is not fit to bind; because, that is to say, the binding of any one with a rope merely *dreamt* of was never witnessed.

c. But, if 'Ignorance' *be* a reality [as some assert], then he declares [as follows]:

न वयं षड्दार्थवादिनो वैशेषिकादिवत् ॥२५॥

A question whether the Vedántí is bound to avoid self-contradiction.	*Aph.* 25 [Possibly the Vedántí may remonstrate] '*We are not asserters of any Six Categories, like the Vaieshikas and others.*'

a. 'We are not asserters of a definite set of categories [like the *Vaieshikas*, who arrange all things under six heads, and the

Naiyáyikas, who arrange them under sixteen]. Therefore, we hold that there *is* such a thing, unknown though it be [to people in general], as 'Ignorance' which is at once real and unreal, or [if you prefer it], which differs at once from the real and the unreal *Vedánta-sára*, § 21]; because this is established by proofs,' [Scriptural or otherwise, which are satisfactory to *us*, although they may not comply with all the technical requisitions of Gotama's scheme of argumentative exposition (see Nyáya Aphorisms, I., § 35)].

b. By the expression [in the aphorism] 'and others' are meant the *Naiyáyikas*; for the *Naiyáyika* is an asserter of sixteen categories [see Nyáya Aphorisms, I., § 1].

c. He confutes [this pretence of evading the objection, by disallowing the categories of the Nyáya]:

न बाह्याभ्यन्तरयोरुपरज्यो परञ्जकभावोऽपि
देशव्यवधानात्सुघ्नस्थपाडलिपुत्रस्थयोरिव ॥२८॥

A thing cannot act where it is not.	Aph. 28 Also [in my opinion, as well as in yours, apparently], between the external and the internal there is not the relation of influenced and influencer; because there is a local separation; as there is between him that stays at Srughna and him that stays at Pátaliputra.

a. In the opinion of these [persons whose theory we are at present objecting to], the soul is circumscribed, residing entirely within the body; and that which is thus *within* cannot stand in the relation of the influenced and the influencer, as regards an *external* object. Why? Because they are separated in regard to place; like two persons the one of whom remains in Srughna (name of a place) and the other in Pátaliputra: such is the meaning. Because the affection which we call 'influence' (*vásaná*) is seen only when there is conjunction, such as that of madder and the cloth [to which it gives its colour], or that of flowers and the flower-basket [to which they impart their odour.]

b. By the word 'also' the absence or conjunction [between the soul and objects (see § *10*)], &c., which he himself holds, is connected [with the matter of the present aphorism]

c. Srughna and Pátaliputra [Palibothra, or Patna] are two several places far apart.

d. But then [these heretics may reply], 'The influence of objects [on the soul] may be asserted, because there is a contact with the object; inasmuch as the soul, according to *us*, goes to the place of the object, just as the senses, according to Your Worship.' Therefore he declares [as follows]:

सत्त्वरजस्तमसां साम्यावस्था प्रकृतिः
प्रकृतेर्महान्महतोऽहंकारोऽहंकारात्मइच्च तन्माचारायुभयमिन्द्रियं तन्मान्नेभ्यः स्थूलभतानि पुरुष इति पञ्चविंशतिर्गराः ॥६॥

| The twenty-five Realitiesenumerated. | Aph. 61 Nature (*prakriti*) is the state of equipoise of Goodness (*sattwa*) Passion (*rajas*), and Darkness (*tamas*): from Nature [proceeds] Mind (*mahat*); from Mind, Self-consciousness (*ahankára*); from Self-consciousness, the five Subtle Elements (*tan-mátra*), and both sets [external and internal,] of Organs (*indriya*); and, from the Subtle Elements, the Gross Elements (*sthúla-bhúta*). [Then there is] Soul(*purusha*). Such is the class of twenty-five. (These things you can see in the Plates as addendum to Sankhya Sutra) |

a. 'The state of equipoise' of the [three] things called 'Goodness,' &c., is their being neither less nor more [one than another]; that is to say, the state of *not* being [developed into] an *effect* [in which one or other of them predominates]. And thus 'Nature' is the triad of 'Qualities' (*guna*), distinct from the products [to which this triad gives rise]: such is the complete meaning.

b. These things, viz., 'Goodness,' &c., [though spoken of as the three Qualities], are not 'Qualities' (*guna*) in the *Vaiseshika* sense of the word; because [the 'Qualities' of the *Vaiseshika* system have, themselves, *no* qualities (see Kanáda's 16th Aph.); while] *these* have the qualities of Conjunction, Disjunction, Lightness, Force, Weight, &c. In this [Sánkhya] system, and in Scripture, &c., the word 'Quality' (*guna*) is employed [as the name of the three things in question], because they are subservient to Soul [and, therefore, hold a secondary rank in the scale of being], and because they form the *cords* [which the word *guna* also signifies], viz., 'Mind,' &c., which consist of the three [so-called] 'Qualities,' and which *bind*, as a [cow, or other] brute-beast, the Soul.

c. Of this [Nature] the principle called 'the great one' (*mahat*), viz., the principle of Understanding, (*buddhi*), is the product. 'Self-consciousness' is a conceit [of separate personality]. Of this there are two products, (1) the p. 73 'Subtile Elements' and (2) the two sets of 'Organs.' The 'SubtileElements' are [those of] Sound, Touch, Colour, Taste, and Smell. The two sets of 'Organs,' through their division into the external and the internal, are of eleven kinds. The products of the 'Subtile Elements' are the five 'Gross Elements.' But 'Soul' is something distinct from either product or cause. Such is the class of twenty-five, the aggregate of things. That is to say, besides these there is nothing.

d. He [next], in [several] aphorisms, declares the order of the inferring [of the existence of these principles, the one from the other]:

स्थूलात्पञ्चज्ञतन्मान्नस्य ॥६२॥

The existence of the 'Subtile Elements' is inferred from that of the 'Gross.'	Aph. 62 [The knowledge of the existence] of the five 'Subtile Elements' is [by inference,] from the 'Gross Elements.'

a. 'The knowledge, by inference,' so much is supplied, [to complete the aphorism, from].

b. Earth, &c., the 'Gross Elements,' are proved to exist, by Perception; [and] thereby [i.e., from that Perception; for Perception must precede Inference, as stated in Gotama's 5th Aphorism,] are the 'Subtle Elements' inferred, [the στοιχεῖαστοιχείων of Empedocles]. And so the application [of the process of inference to the case] is as follows:

(1) The Gross Elements, or those which have not reached the absolute limit [of simplification, or of the atomic], consist of things [Subtle Elements, or Atoms,] which have distinct qualities; [the earthy element having the distinctive quality of Odour; and so of the others]:

(2) Because they are gross;

(3) [And everything that is gross is formed of something less gross, or, in other words, more subtle,] as jars, webs, &c.; [the gross web being formed of the less gross threads; and so of the others].

तेनान्तःकररास्य ॥६४॥

And thence that of Intellect.	Aph. 64 [The knowledge of the existence] of Intellect is [by inference,] from that [Self-consciousness,]

a. That is to say: by inference from [the existence of] 'that,' viz., Self-consciousness, which is a product, there comes the knowledge of 'Intellect' (*buddhi*), the *great* 'inner organ' (*antahkarana*), [hence] called 'the great one' (*mahat*), [the existence of which is recognized] under the character of the *cause* of this [product, viz., Self-consciousness].

b. And so the application [again rather circular, of the process of inference to the case,] is as follows:

(1) The thing called Self-consciousness is made out of the things that consist of the moods of judgment [or mind];

(2) Because it is a thing which is a product of judgment [proceeding in the Cartesian order of *cogito, ergo sum*; and]

(3) Whatever is not so [i.e., whatever is *not* made out of judgment, or mental assurance], is not thus [i.e., is not a product of mental assurance]; as the Soul, [which is not made out of this or of anything antecedent], &c.

c. Here the following reasoning is to be understood: Every one, having first determined anything under a concept [i.e., under such a form of thought as is expressed by a general term; for example, that this which presents itself is a jar, or a human body, or a possible action of one kind or other], after that makes the judgment, 'This is I,' or 'This ought to be done by me,' and so forth: so much is quite settled; [and there is no dispute that the fact is as here stated]. Now, having, in the present instance, to look for some *cause* of the thing called 'Self-consciousness' [which manifests itself in the various judgments just referred to], since the relation of cause and effect subsists between the two functions [the occasional conception, and the subsequent occasional judgment, which is a function of Self-consciousness], it is assumed, for simplicity, merely that the relation of cause and effect existsbetween the two substrata to which the [two sets of] functions belong; [and this is sufficient,] because it follows, as a matter of course, that the occurrence of a *function* of the effect must result from the occurrence of a *function* of the cause; [nothing, according to the Sánkhya, being in any product, except so far, and in such wise, as it preexisted in the cause of that product].

ततः प्रकृतः ॥६५॥

| And thence that of Nature. | Aph. 65 [The knowledge of the existence] of Nature is [by inference,] from that ['Intellect,']. |

a. By inference from [the existence of] 'that,' viz., the principle [of Intellect, termed], 'the Great one,' which is a *product*, there comes the knowledge of [the existence of] Nature, as [its] cause.

b. The application [of the process of inference to the case] is as follows:

(1) Intellect, the affections whereof are Pleasure, Pain, and Dulness, is produced from something which has these affections, [those of] Pleasure, Pain, and Dulness:

(2) Because, whilst it is a *product* [and must, therefore, have arisen from something consisting of that which itself now consists of], it consists of Pleasure, Pain, and Dulness; [and]

(3) [Every *product* that has the affections of, or that occasions, Pleasure, Pain, or Dulness, takes its rise in something which consists of these]; as lovely women, &c.

c. For an agreeable woman gives pleasure to her husband, and, therefore, [is known to be mainly made up of, or] partakes of the quality of 'Goodness;' the indiscreet one gives pain to him, and, therefore, partakes of the quality of 'Foulness;' and she who is separated [and perhaps forgotten,] occasions indifference, and so partakes of the quality of Darkness.'

d. And the appropriate refutation [of any objection], in this case, is [the principle], that it is fitting that the qualities of the effect should be [in every case,] in conformity with the qualities of the cause.

e. Now he states how, in a different way, we have [the evidence of] inference for [the existence of] Soul, which is void of the relation of cause and effect that has been mentioned, [in the four preceding aphorisms, as existing between Nature and its various products]:

संहतपरार्थत्वात्पुरुषस्य ॥६६॥

The argument for the existence of Soul.	Aph. 66 [The existence] of Soul [is inferred] from the fact that the combination [of the principles of Nature into their various effects] is for the sake of another [than unintelligent Nature, or any of its similarly unintelligent products].

a. 'Combination,' i.e., conjunction, which is the cause [of all products; these resulting from the conjunction of their constituent parts]. Since whatever has this quality, as Nature, Mind, and so on [unlike soul, which is *not* made up of parts], is for the sake of some other; for this reason it is understood that soul exists: such is the remainder, [required to complete the aphorism].
b. But the application [of the argument, in this particular case, is as follows]:
(1) The thing in question, viz., Nature the 'Great one,' with the rest [of the aggregate of the unintelligent], has, as its fruit [or end], the [mundane] experiences and the [eventual] Liberation of some other than itself:
(2) Because it is a combination [or *compages*];
(3) [And every combination,] as a couch, or a seat, or the like, [is for another's use, not for its own; and its several component parts render no mutual service].
c. Now, in order to establish that it is the cause of all [products], he establishes the *eternity* of Nature (*prakriti*):

मूले मूलाभावादमूलं मूलम् ॥६७॥

| Argument for the eternity of Nature. | Aph. 67 Since the root has no root, the root [of all] is rootless. |

a. Since 'the root' (*múla*), i.e., the cause of the twenty three principles, [which, with soul and the root itself, make up the twenty-five realities recognized in the Sánkhya,] 'has no root,' i.e., has no cause, the 'root,' viz., Nature (*pradhána*), is 'rootless,' i.e., void of root. That is to say, there is no other cause of Nature; because there would be a *regressus in infinitum*, [if we were to suppose another cause, which, by parity of reasoning, would require another cause; and so on without end].
b. He states the argument [just mentioned] in regard to this, [as follows]:

Sāmkhya/Sānkhya Sutra Kapila

श्रधिकारन्नैविध्यान्न नियः ॥७०॥

All do not profit by the saving truth; because it is only the best kind of people that are fully amenable to reason.	Aph. 70 There is no rule [or necessity, that *all* should arrive at the truth]; because those who are privileged [to engage in the inquiry] are of three descriptions.

a. For those privileged [to enguge in the inquiry] are of three descriptions, through their distinction into those who, in reflecting, are dull, mediocre, and best. Of these, by the dull the [Sánkhya] arguments are frustrated [and altogether set aside], by means of the sophisms that have been uttered by the *Bauddhas*, &c. By the mediocre they [are brought into doubt, or, in other words,] are made to appear as if there were equally strong arguments on the other side, by means of arguments which really prove the reverse [of what these people employ them to prove], or by arguments which are not true: [see the section on Fallacies in the *Tarka-sangraha*]. But it is only the *best* of those privileged, that reflect in the manner that has been set forth [in our exposition of the process of reflexion which leads to the discriminating of Soul from Nature]: such is the import. But there is no rule that *all* must needs reflect in the manner so set forth: such is the literal meaning.

b. He now, through two aphorisms, defines 'the Great one' and 'Self-consciousness'; [the reader being presumed to remember that Nature consists of the three 'Qualities' in equipoise, and to be familiar with the other principles, such as the 'Subtile elements']:

महदाख्यमाद्यं कार्यं तन्मनः ॥७१॥

By 'the Great one' is meant Mind.	Aph. 71 The first product [of the Primal Agent, Nature], which is called 'the Great one,' is Mind.

a. 'Mind' (*manas*). 'Mind' [is so called], because its function is 'thinking' (*manana*). By 'thinking' is here meant 'judging' (*nischaya*). That of which this is the function is 'intellect' (*buddhi*); and *that* is the first product, that called 'the Great one' (*mahat*): such is the meaning.

परिच्छन्नं न सर्वोपादानम् ॥७६॥

Why the theory of a plastic Nature is preferable to that of Atoms.	Aph. 76 What is limited cannot be the substance of all [things].

a. That which is limited cannot be the substance of all [things]; as yarn cannot be the [material] cause of a jar. Therefore it would [on the theory suggested,] be necessary to mention separate causes of [all] things severally; and it is simpler to assume a single cause. Therefore Nature alone is the cause. Such is the meaning.
b. He alleges Scripture in support of this:

तदुत्मन्तिश्रुतेश्च ॥७७॥

Scripture declares in favour of the theory.	Aph. 77 And [the proposition that Nature is the cause of all is proved] from the text of Scripture, that the origin [of the world] is therefrom, [i.e., from Nature].

a. An argument, in the first instance, has been set forth [for, till argument fails him, no one falls back upon authority]. Scripture, moreover, declares that Nature is the cause of the world, in such terms as, 'From Nature the world arises,' &c.
b. But then [some one may say], a jar which antecedently did not exist is seen to come into existence. Let, then, *antecedent non-existence* be the cause [of each product]; since this is an invariable antecedent, [and, hence, a cause; 'the invariable antecedent being denominated a cause,' if Dr. Brown, in his 6th lecture, is to be trusted]. To this he replies:

नावस्तुनो वस्तुसिध्दः ॥७८॥

| Ex nihilo nihil fit. | Aph. 78 A thing is not made out of nothing. |

- a. That is to say: it is not possible that out of nothing, i.e., out of a nonentity, a thing should be made, i.e., an entity should arise. If an entity were to arise out of a nonentity, then, since the character of a cause is visible in its product, the *world*, also, would be unreal: such is the meaning.
- b. Let the world, too, *be* unreal: what harm is that to us? [If any ask this,] he, therefore, declares [as follows]:

श्रबाधाददुष्टकाररजन्यात्वाच्च नावस्तुत्वम् ॥७९॥

| Reasons why the world is not to be supposed unreal. | Aph. 79 It [the world] is not unreal; because there is no fact contradictory [to its reality], and because it is not the [false] result of depraved causes, [leading to a belief in what ought not to be believed]. |

- a. When there is the notion, in regard to a shell [of a pearl-oyster, which sometimes glitters like silver], that it is silver, its being silver is contradicted by the [subsequent and more correct] cognition, that this is *not* silver. But, in the case in question [that of the world regarded as a reality], no one ever has the cognition, 'This world is *not* in the shape of an entity,' by which [cognition, if any one ever really had such,] it being an entity might be opposed.
- b. And it is held that that is false which is the result of a *depraved* cause; e.g., some one's cognition of a [white] conch-shell as *yellow*, through such a fault as the jaundice, [which depraves his eye-sight]. But, in the case in question, [that of the world regarded as a reality], there is not such [temporary or occasional] depravation [of the senses]; because all, at all times, cognize the world as a reality. Therefore the world is *not* an unreality.

c. But then [someone may suggest], *let* a nonentity be the [substantial] cause of the world; still the world will not [necessarily, therefore,] be unreal. In regard to this, he declares [as follows]:

त्रिविधं प्रमाणं तत्सिद्धौ सर्वसिद्धेर्नाधिक्यसिद्धिः ॥८८॥

There are three kinds of evidence.	Aph. 88 Proof is of three kinds there is no establishment of more; because, if these be established, then all [that is true] can be established [by one or other of these three proofs].

a. 'Proof is of three kinds;' that is to say, 'perception' (*pratyaksha*), 'the recognition of signs' (*anumána*), and 'testimony' (*sabda*), are the [three kinds of] proofs.

b. But then [some one may incline to say], let 'comparison' [which is reckoned, in the Nyáya, a specifically distinct source of knowledge], and the others [such as 'Conjecture,' &c., which are reckoned, in like manner, in the Mímánsá], also be instruments of right knowledge, [as well as these three], in [the matter of] the discriminating of Nature and Soul: he therefore says, 'because, if those [three] be established,' &c. And, since, if there be the three kinds of proof established, 'everything [that is really true] can be established [by means of them], there is no establishment of more;' no addition to the proofs can be fairly made out; because of the cumbrousness [that sins against the philosophical maxim, that we are not to assume more than is necessary to account for the case]: such is the meaning.

c. For the same reason, Manu, also, has laid down only a triad of proofs, where he says. 'By that man who seeks a distinct knowledge of his duty, [these] three [sources of right knowledge] must be well understood, viz., Perception, Inference, and Scriptural authority in its various shapes [of legal institute, p. 108 &c.].' And 'Comparison,' and 'Tradition' (*aitihya*), and the like, are included under Inference and Testimony; and 'Non-perception' (*anupalabdhi*) and the like are included under Perception; [for the non-perception of an absent jar on a particular spot of ground is nothing else than the perception of that spot of ground *without* a jar on it].

d. He [next] states the definitions of the varieties [of proof, having already given the general definition]:

ईश्वरासिद्धेः ॥९२॥

| That any 'Lord' exists is not proved. | Aph. 92 [This objection to the definition of Perception has no force]; because it is not proved that there *is* a Lord (*íswara*). |

a. That there is no fault [in the definition of Perception], because there is no proof that there *is* a Lord, is supplied.

b. And this demurring to there being any 'Lord' is merely in accordance with the arrogant dictum of [certain] partisans [who hold an opinion not recognized by the majority]. Therefore, it is to be understood, the expression employed is, 'because it is *not proved* that there is a Lord,' but not the expression, 'because there *is no* Lord.'

c. But, on the implication that there *is* a 'Lord,' what we mean to speak of [in our definition of Perception,] is merely the being of the [same] kind with what is produced by conjunction [of a sense-organ with its object; and the perceptions of the 'Lord' may be of the same *kind* with such perceptions, though they were not to come from the same *source*].

d. Having pondered the doubt, '*How* should the Lord not be proved [to exist] by the Scripture and the Law, [which declare his existence]?' he states a dilemma which excludes [this]:

मुक्तबद्धयोरन्यतराभावान्न तत्सिद्धिः ॥९३॥

| A dilemma, to exclude proof that there is any 'Lord.' | Aph. 93 [And, further,] it is not proved that he [the 'Lord,'] exists; because [whoever exists must be either free or bound; and], of free and bound, he can be neither the one nor the other. |

Creation of Universe – God or Big Bang

a. The 'Lord' whom you imagine, tell us, is he free from troubles, &c.? Or is he in bondage through these? Since he is not, cannot be, either the one or the other, it is not proved that there is a 'Lord:' such is the meaning.
b. He explains this very point:

उभयथाप्यसत्करत्वम् ॥९४॥

The force of the dilemma.	Aph. 94 [Because,] either way, he would be inefficient.

a. Since, if he were free, he would have no desires, &c., which [as compulsory motives,] would instigate him to create; and, if he were bound, he would be under delusion; he must be [on either alternative,] unequal to the creation, &c. [of this world].
b. But then, [it may be asked,] if such be the case, what becomes of the Scripture-texts which declare the 'Lord?' To this he replies:

परिमाणात् ॥१३०॥

A second proof.	Aph. 130 Because of [their] measure, [which is a limited one].

a. That is to say: [Mind and the rest are products]; because they are limited in measure; [whereas the only two that are uncaused, viz., Nature and Soul, are unlimited].
b. He states another argument:

शक्तितश्चेति ॥१३२॥

A fourth proof.	Aph. 132 And, finally, because it is through the power [of the cause alone, that the product can do aught].

a. It is by the power of its cause, that a product energizes, [as a chain restrains an elephant, only by the force of the iron which it is made of]; so that Mind and the rest, being [except through the strength of Nature,] powerless, produce *their* products in subservience to Nature. Otherwise, since it is their habit to energize, they would at all times produce their products, [which it will not be alleged that they do].

b. And the word *iti*, in this place, is intended to notify the completion of the set of [positive] reasons [why Mind and the others should be regarded as products].

c. He [next] states [in support of the same assertion,] the argument from negatives, [i.e., the argument drawn from the consideration as to what becomes of Mind and the others, when they are *not* products]:

कार्यात्कारणानुमानं तत्साहित्यात् ॥१३५॥

What kind of causes can be inferred from their effects.	Aph. 135 The cause is inferred from the effect, [in the case of Nature and her products]; because it accompanies it.

a. That [other relation, other than that of material and product, which you would make out to exist between Nature and Mind,] exists, indeed, where the nature of the cause is not seen in the effect; as [is the case with] the inference, from the rising of the moon, that the sea is swollen [into full tide; rising, with maternal affection, towards her son who was produced from her bosom on the occasion of the celebrated Churning of the Ocean. Though the swelling of the tide does not occur apart from the rising of the moon, yet here the cause, moon-rise, is not seen in the effect, tide; and, consequently, though we infer the effect from the cause, the cause could not have been inferred from the effect]. But, in the present case, since we see, in Mind and the rest, the characters of Nature, the cause *is* inferred from the effect. 'Because it accompanies it,' i.e., because, in

b. Mind and the rest, we see the properties of Nature, [i.e., Nature herself actually present; as we see the clay which is the cause of a jar, actually present in the jar].

c. [But it may still be objected,] if it be thus, then let that principle itself, the 'Great one' [or Mind], be the cause of the world: what need of *Nature*? To this he replies:

शरीरादिव्यतिरिक्तः पुमान् ॥१३९॥

Materialism scouted.	Aph. 139 Soul is something else than the body, &c.

a. [The meaning of the aphorism is] plain.
b. He propounds an argument in support of this:

त्रिगुणादिविपर्ययात् ॥१४१॥

Soul presents no indication of being material.	Aph. 141 [And Soul is something else than the body, &c.]; because there is [in Soul,] the reverse of the three Qualities, &c.

a. Because there is, in Soul, 'the reverse of the three Qualities,' &c., i.e., because they are not seen [in it]. By the expression '&c.' is meant, because the *other* characters of Nature, also, are not seen [in soul].
b. He states another argument:

श्राधिष्ठानाच्चेति ॥१४२॥

Another proof that Soul is not material.	Aph. 142 And [Soul is not material;] because of [its] superintedence [over Nature].

a. For a superintendent is an intelligent being; and Nature is unintelligent: such is the meaning.
b. He states another argument:

भोक्तृभावात् ॥१४३॥

Another proof.	Aph. 143 [And Soul is not material;] because of [its] being the experiencer.

a. It is Nature that is experienced; the experiencer is Soul. Although Soul, from its being unchangeably the same, is not [really] an experiencer, still the assertion [in the aphorism,] is made, because of the fact that the reflexion of the Intellect befalls it, [and thus makes it *seem* as if it experienced].

b. Efforts are engaged in for the sake of Liberation. Pray, is this [for the benefit] of the Soul, or of Nature; [since Nature, in the shape of Mind, is, it seems, the experiencer]? To this be replies:

सुषुप्त्यचाध्यसाश्रित्वम् ॥१४८॥

Argument against the soul's being unintelligent.	Aph. 148 [If soul were unintelligent,] it would not be witness [of its own comfort,] in profound [and dreamless] sleep, &c.

a. If soul were unintelligent, then, in deep sleep, &c., it would not be a witness, a knower. But that this is not the case [may be inferred] from the phenomenon, that 'I slept *pleasanty*.' By the expression '&c.' [in the aphorism,] dreaming is included.

b. The Vedántís say that 'soul is *one* only'; and so, again, 'For Soul is eternal, omnipresent, changeless, void of blemish.' 'Being one [only], it is divided [into a seeming multitude] by Nature (*śakti*), i.e., Illusion (*máyá*), but not through its own essence, [to which there does not belong multiplicity].' In regard to this, he says [as follows]:

जन्मादिव्यवस्थातः पुरुषबहुत्वम् ॥१४९॥

There is a multiplicity of souls.	Aph. 149 From the several allotment of birth, &c., a multiplicity of souls [is to be inferred].

a. 'Birth, &c.' By the '&c.,' growth, death, &c., are included. 'From the several allotment' of these, i.e., from their being appointed; [birth to one, death to another, and so on]. 'A multiplicity of souls;' that is to say, souls are many. If soul were one only, then, when *one* is born, *all* must be born, &c. Vedanta and Upanishads uphold that soul is one. Soul is present in every human body. Upon death soul gets liberated and becones one with One which is undivisible)
b. He ponders, as a doubt, the opinion of the others, [viz., of the Vedántís]:

नान्धादष्टच्चा चक्षष्मतामनुपलम्भः ॥१५६॥

| He jeers at the Vedántí | Aph. 156 No: because the blind do not see, can those who have their eyesight not perceive? |

a. What! because a blind man does not see, does also one who has his eyesight not perceive? There are *many* arguments [in support of the view] of those who assert that souls are many, [though *you* do not see them]: such is the meaning.'
b. He declares, for the following reason, also, that souls are many:

वामदेनादिर्मुक्तो नाध्दैतम् ॥१५७॥

| Scripture proof that Souls are many. | Aph. 157 Vámadeva, as well as others, has been liberated, [if we are to believe the Scriptures; therefore] non-duality is not [asserted, in the same Scriptures, in the Vedántic sense]. |

a. In the Puránas, &c., we hear, 'Vámadeva has been liberated,' 'Suka has been liberated,' and so on. If Soul were *one*, since the liberation of all would take place, on the liberation of one, the Scriptural mention of a diversity [of separate and successive liberations] would be self-contradictory.
b. [But the Vedántí may rejoin:] on the theory that Souls are many, since the world has been from eternity, and from time to time some

one or other is liberated, so, by degrees, *all* having been liberated, there would be a universal void. But, on the theory that Soul is *one*, Liberation is merely the departure of an adjunct, [which, the Vedántí flatters himself, does not involve the inconsistency which he objects to the Sánkhya], To this he replies:

Sankhya Sutra Aphorisms Book V

स्वोपकारादधिष्ठानं लोकवत् ॥३॥

| The supposed Lord would be selfish. | Aph. 3. [If a Lord were governor, then] from intending his own benefit, his government [would be selfish], as is the case [with ordinary governors] in the world. |

a. If the Lord were the governor, then his government would be only for his own benefit; as is the case [the case with ordinary rulers] in the world: such is the meaning.[4]
b. In reply to the doubt, 'grant that the Lord, also, be benefited: what harm?' he says:[1]

लौकिकेश्वरवदितरस्या ॥४॥

| And, therefore, not the Lord spoken of. | Aph. 4. [He must, then, be] just like a worldly lord, [and] otherwise [than you desire that we should conceive of him]. |

a. If we agree that the Lord, also, is benefited, he, also, must be something mundane, 'just like a worldly lord;' because, since his desires are [on that supposition,] not [previously] satisfied, he must be liable to grief, &c.: such is the meaning.[2]
b. In reply to the doubt, 'be it even so,' he says:[3]

पारिभाषिको वा ॥५॥

| The difficulty perhaps originates in a mistaken expression. | Aph. 5. Or [let the name of Lord be] technical. |

a. If, whilst there exists also a world, there be a Lord, then let yours, like ours, be merely a technical term for that soul which emerged at the commencement of the creation; since there cannot be an eternal lordship, because of the contradiction between mundaneness and the having an unobstructed will: such is the meaning.¹

b. He states another objection to the Lord's being the governor:³

न रागाद्ते ³ तत्सिङ्गचक्रः प्रतिनियतकारस्रात्वात् ॥६॥

Objection to there being a Lord.	Aph. 6. This [position, viz., that there is a Lord,] cannot be established without [assuming that he is affected by] Passion; because that is the determinate cause [of all energizing].

a. That is to say: moreover, it cannot be proved that he is a *governor*, unless there be Passion; because Passion is the determinate cause of activity.⁴

b. But then, be it so, that there is Passion in the Lord, even. To this he replies:¹

तथोगेडपि न नित्यमुक्तः ॥७॥

This objection, further.	Aph. 7. Moreover, were that [Pasion] conjoined with him, he could not be eternally free.

a. That is to say: moreover, if it be agreed that there is conjunction [of the Lord] with Passion, he cannot be eternally free; and, therefore, thy tenet [of his eternal freedom] is invalidated.²

b. Pray [let us ask], does lordship arise from the immediate union, with Soul, of the wishes, &c., which we hold to be properties of Nature, [not properties of Soul]? Or from an influence by reason of the mere existence of proximity, as in the case of the magnet? Of these he condemns the former alternative:³

प्रधानशक्तियोगाच्चेत्सङ्गापत्तिः:¹ ॥८॥

| Objection, on one branch of an alternative. | *Aph.* 8. If it were from the conjunction of the properties of Nature, it would turn out that there is association, [which Scripture denies of Soul]. |

a. From the conjunction, with Soul, of 'the properties of Nature,' i.e., Desire, &c., Soul, also, would turn out [contrary to Scripture,] to be associated with properties.²

b. But, in regard to the latter [alternative], he says:³

सत्तामात्राच्चेत्सर्वैश्वर्यम् ॥९॥

| Objection, on the other branch. | *Aph.* 9. If it were from the mere existence [of Nature, not in association, but simply in proximity], then lordship would belong to every one. |

a. That is to say: if lordship is by reason of the mere existence of proximity, as in the case of the magnet [which becomes affected by the simple proximity of iron], then it is settled, as we quite intend it should be, that even all men, indifferently, experiencers in this or that [cycle of] creation, [may] have lordship; because it is only by conjunction with all experiencers, that Nature produces Mind, &c. And, therefore, your tenet of there being only one Lord is invalidated.¹

b. Be it as you allege; yet these are false reasonings; because they contradict the evidence which establishes [the existence of] a Lord. Otherwise, Nature, also, could be disproved by thousands of false reasonings of the like sort. He therefore says:²

प्रमाण प्रमाणाभावान्न तत्सिद्धिः ॥१०॥

| Denial that there is any evidence of a Lord. | *Aph.* 10. It is not established [that there is an eternal Lord]; because there is no evidence of it. |

a. Its establishment, i.e., the establishing that there is an eternal Lord. Of the Lord, in the first place, there is not *sense*-evidence; so that only the evidences of inference and of testimony can be offered; and these are inapplicable: such is the meaning.[1]

b. The inapplicability he sets forth in two aphorisms:[2]

संबन्धाभावान्नानुमानम् ॥ ९ ॥

Denial that it can be established by inference.	Aph. 11. There is no inferential proof [of there being a Lord]; because there is [here] no [case of invariable] association [between a sign and that which it might betoken].

a. 'Association,' i.e., invariable concomitancy. 'There is none;' i.e., none exists, [in this case]. And so there is no inferential proof of there being a Lord; because, in such arguments as, 'Mind, or the like, has a maker, because it is a product,' [the fact of] invariable concomitancy[3] is not established, since there is no compulsion [that every product should have had an intelligent maker]. Such is the meaning.[4]

Creation as per Sankhya Sutra represented in pictoral form

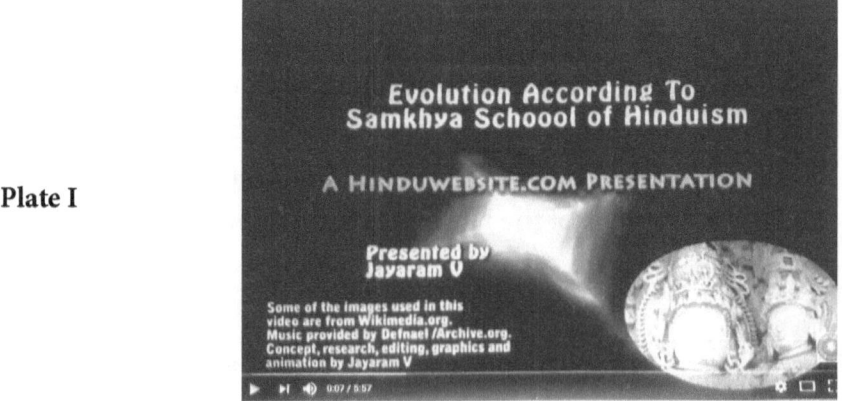

Plate I

Plate II

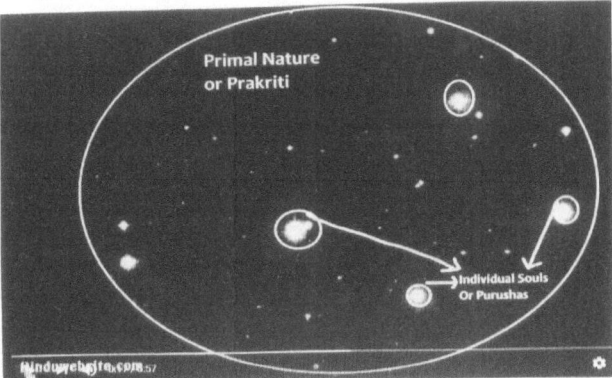

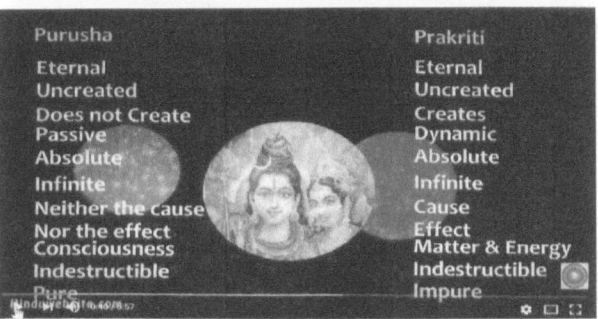

Plate III

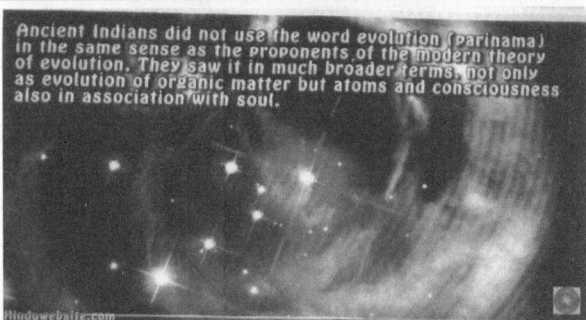

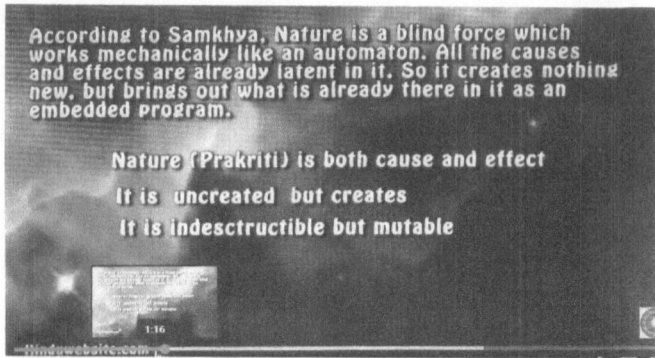

Plate IV

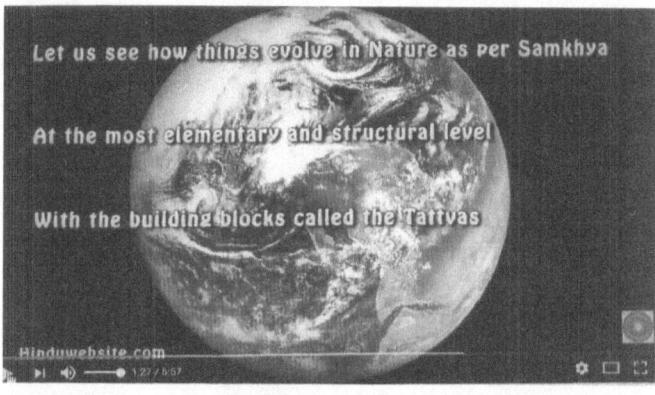

Plate V

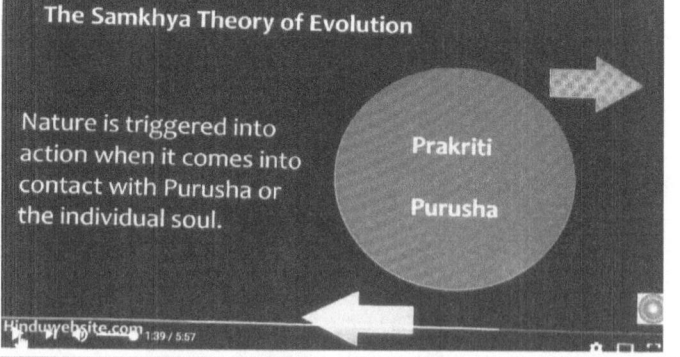

Sāmkhya/Sānkhya Sutra Kapila

Plate VI

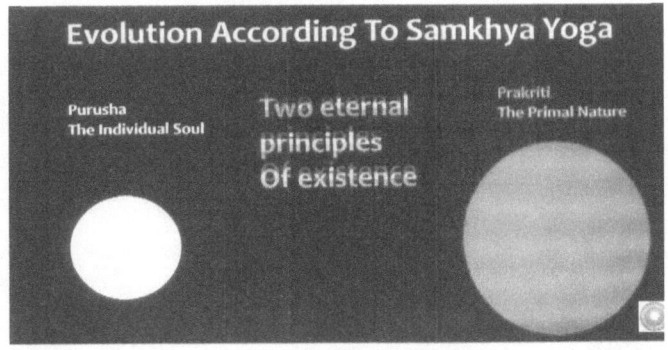

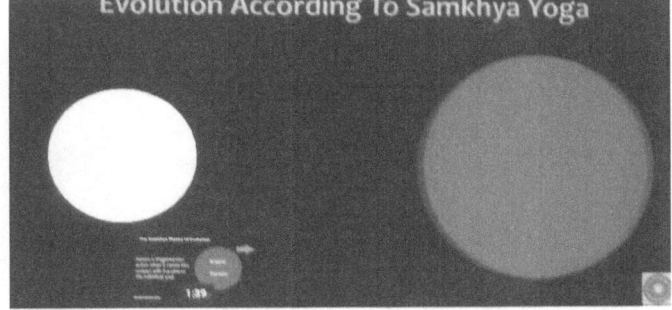

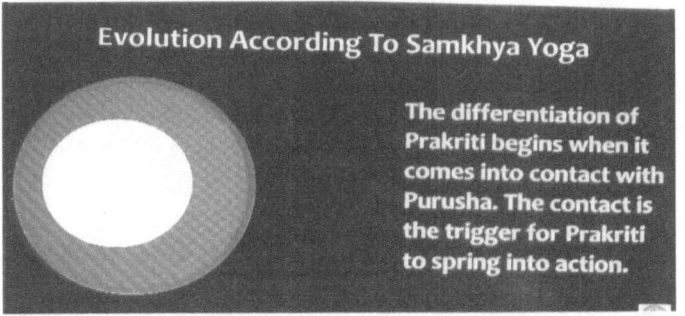

Plate VII

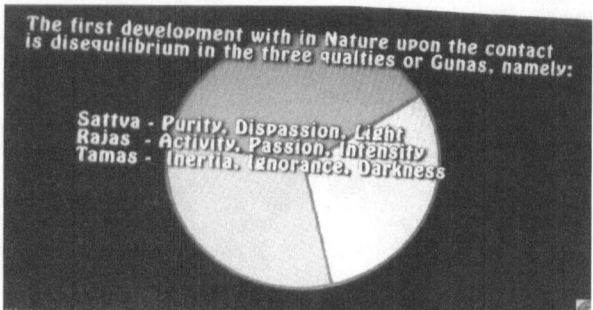

Plate XII

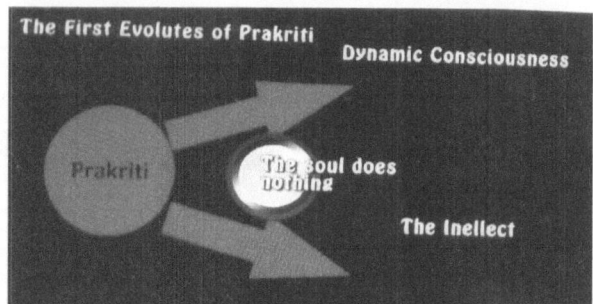

Plate XIII

Plate VIII

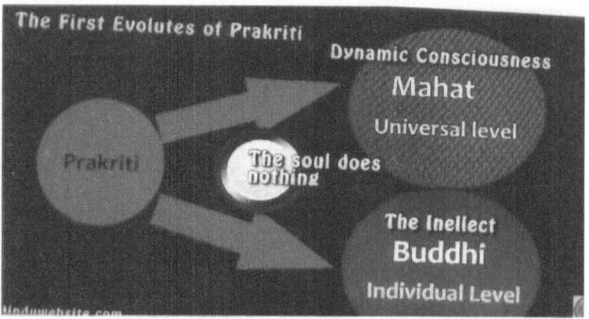

Plate XIV

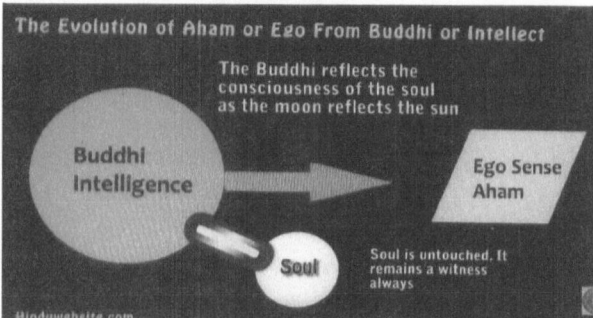

Plate XV

Plate IX

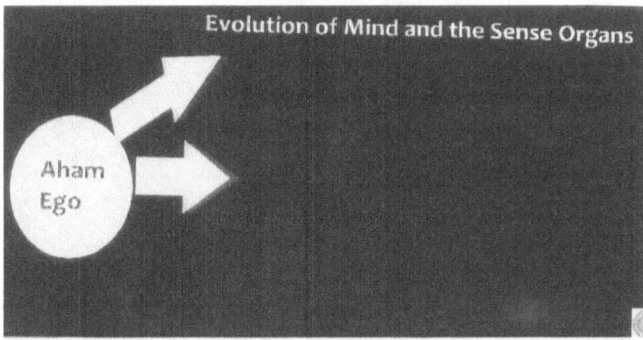

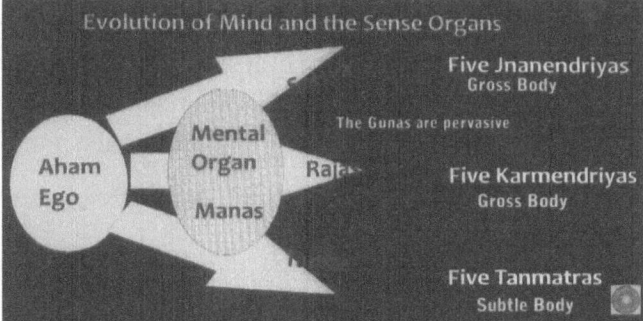

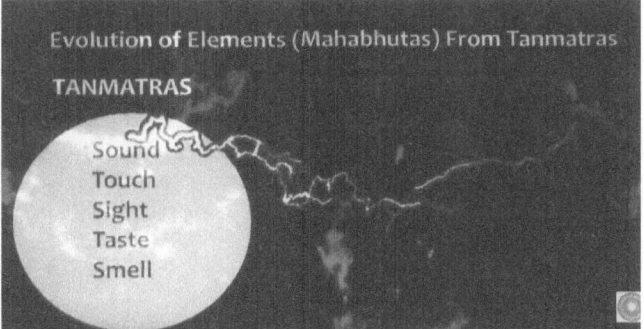

Plate X

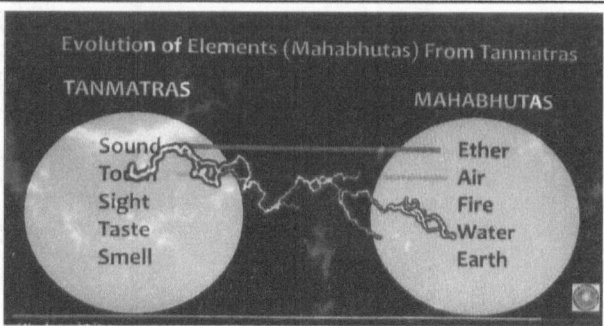

Plate XIX

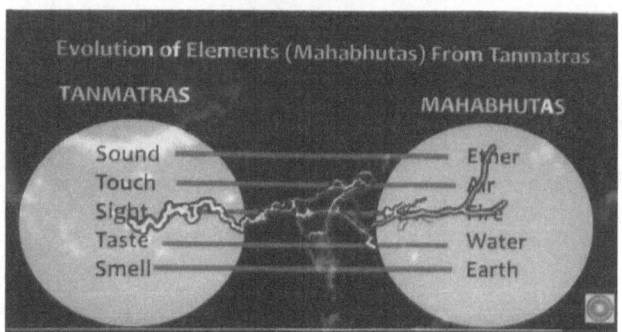

Plate XX

Plate XI

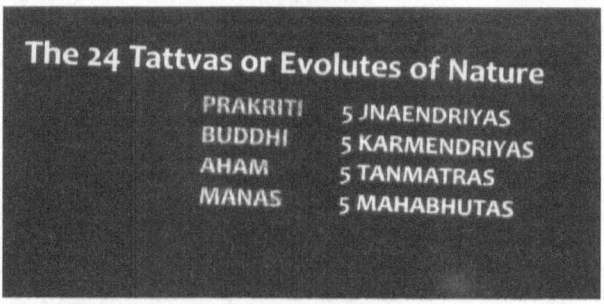

Plate XXI

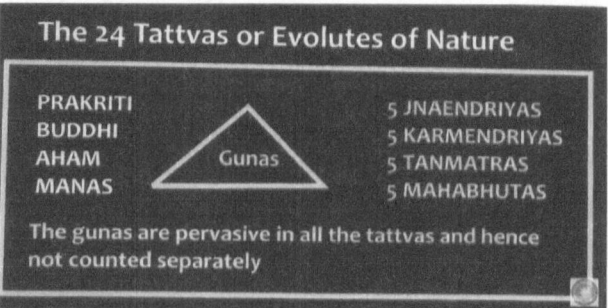

Sāmkhya/Sānkhya Sutra Kapila

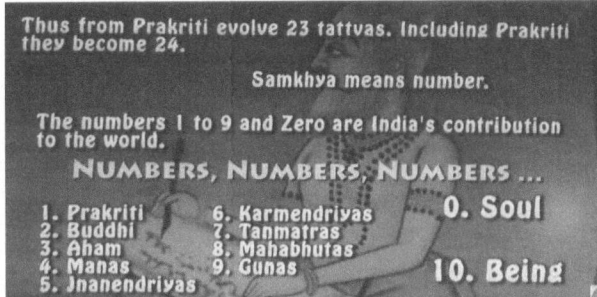

Plate XXIII

Plate XII

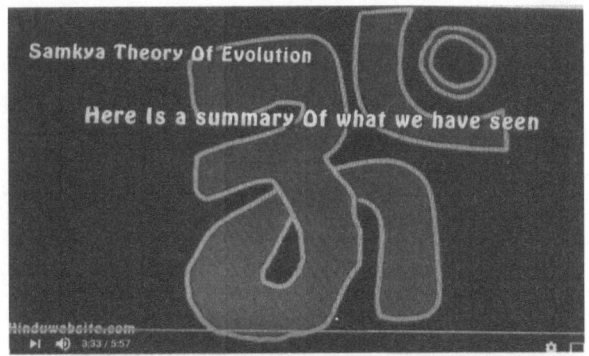

Plate XIII

Plate XXV

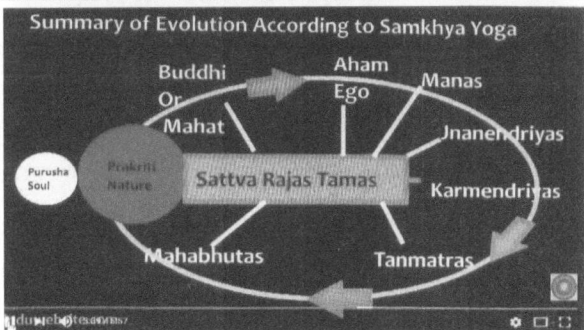

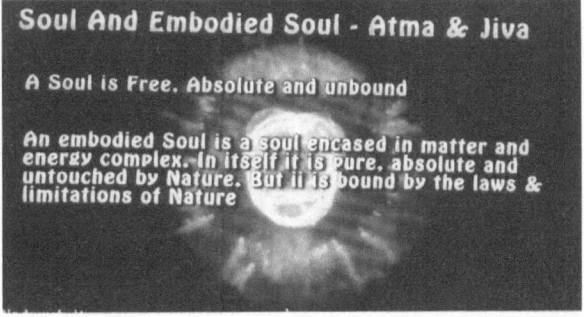

Plate XIV

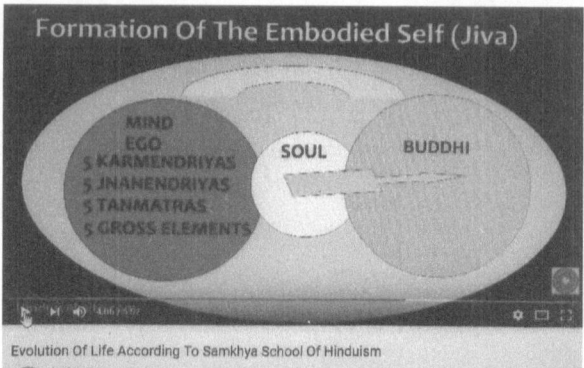

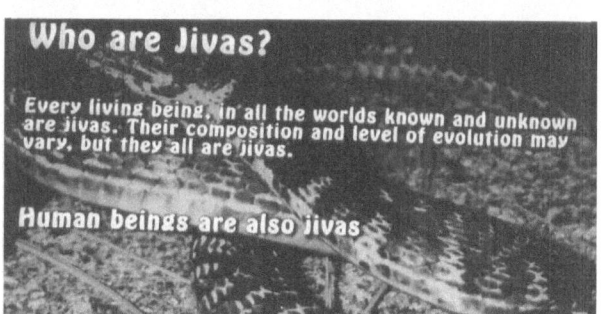

Plate XV

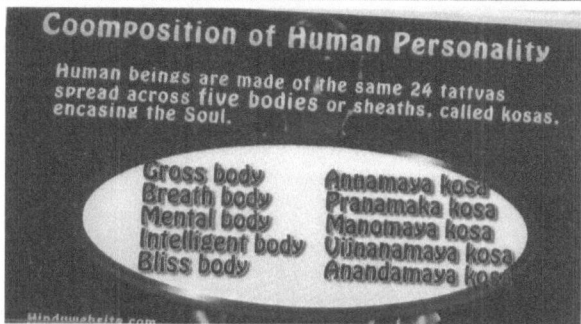

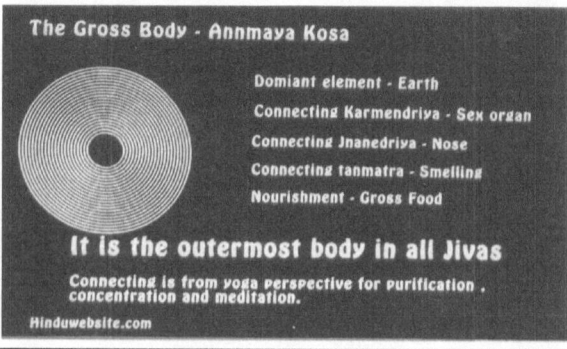

Plate XVI

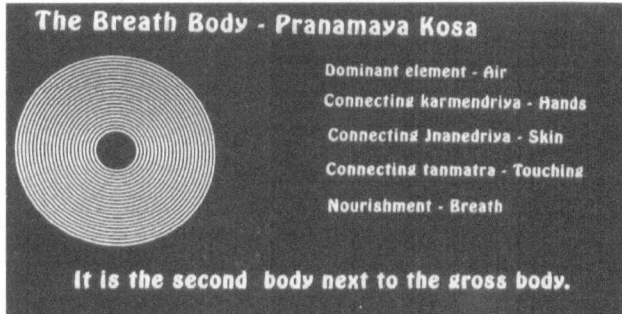

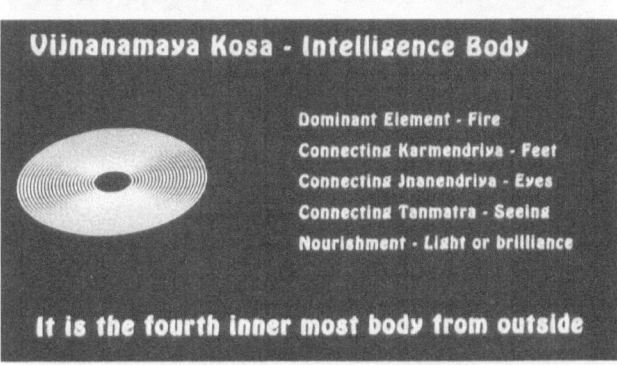

Plate XVII

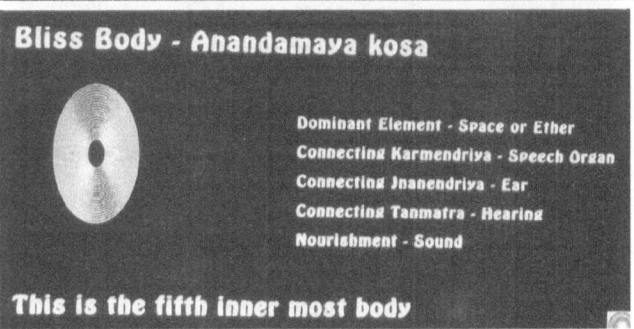

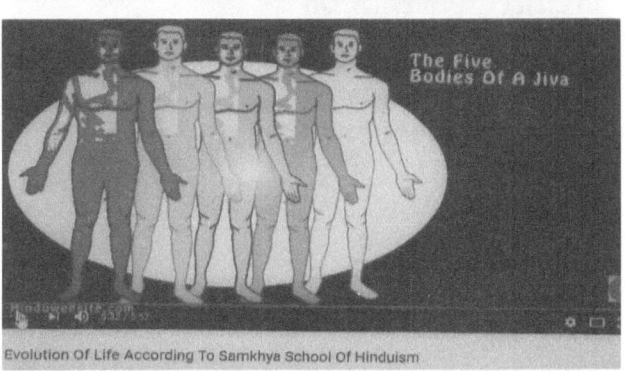

CHAPTER 5
CREATION AS PER RIG VEDA

by Pujya Dr. RL Kashyap

INTRODUCTION

The Rigveda (Sanskrit: ऋग्वेद rgveda, from [rc "Praise, Shine" [1] and veda "Knowledge") is an ancient collection of vedic Sanskrit hymns. It is one of the four and oldest Canonical sacred texts (Sruti) of Hinduism known as the Vedas.

Creation of universe in Rigveda is given in 10th Mandala of Rig Veda.

R V: X.121.1

हिरण्यगर्भः समवर्तताग्रे
भूतस्ये जातः पतिरेक आसीत्
स दाधार पृथिवीं द्यामुतेमां
कस्मै देवाय हविषा विधेम

य आत्मदा बलदा यस्य
विश्वे उपासते प्रशिषं यस्य देवाः
यस्य छायामृतं यस्य मृत्युः
कस्मै देवाय हविषा विधेम

यः प्राणतो निमिषतो
महित्वा
एक इद्राजा
जगते बभूव

Creation as per Rig Veda

यईशे अस्य द्विपदश्चतुष्पदः
कस्मै देवाय हविषा विधेम

R V: X.121.1: In the beginning (agre) arose the golden seed (1);
   born, he was the solelord of every creature (2).
   He upheld this earth and heaven (3).
   Which deva (except him) shall we worship with
      offerings (4)?

[*hiranya*: gold; note that the common meaning is gold. But *hi* stands for *hita*, 'placed' or 'hidden' and *ranya* means 'delight' in many RV mantras. Hence *hiranya* is that 'in which the delight is hidden.' It is the concrete image of the higher light, the gold of the Truth. Gold is the symbolic colour of the light of the Sun (SA). *kah*: the pronoun 'who'; Prajapati. Many translations of the line (4) read, who is the diety we shall worship?', reflecting a doubt. The sages had no such doubts. They were struck by a sense of wonder at the entire creation. (R V: X.121.10) clearly states that Prajapati is the creator. *vidhema*: worship]

R V: X.121.2: It is he who bestows the soul-force (*atmada*) and vigour (1);
   upon his call; all, even Gods, approach (him) (2).
   His shadow is immortality; death is also his (shadow) (3).
   Which *deva* (except him) shall we worship with
      offerinds (4)?

[*bala*: vigour; *prasisha*: call; *upasate*: approach (upa + asate)])

R V: X.121.3: In this Universe he by his greatness became (2, 4),
   the sole (*eka*) king of the breathing and seeing (1, 3).
   He is the lord of all beings with two states and four (3).
   Which *deva* (except him) shall we worship
      with offerings (4)?

[*dvipade*: two states of consciousness such as giving and taking; matter and mind; here and hereafter etc.; *chatushpade*: four states; matter, life, mind and soar (super-mind)]

यस्येमे हिमवन्तो महित्वा
यस्य समुद्रं रसया सहाहुः
यस्येमाः प्रदिशो यस्य बाहू
कस्मै देवाय हविषा विधेम

येन द्यौरुग्रा पृथिवी चे दृळ्हा
अभ्यैक्षेतां मनसा रेजमाने
यत्राधि सूर उदितो विभाति
कस्मै देवाय हविषा विधेम

यं क्रन्दसी
अवसा तस्तभाने
अभ्यैक्षेतां सनसा रेजमाने
यत्राधि सूर उदितो विभाति
कस्मै देवाय हविषा विधेम

आपो ह यत् बृहतीः विश्रमायन्
गर्भं दधाना जनयन्तीः अग्निम्
ततों देवाना समवर्तत असः एकछ
कस्मै देवाय हविषा विधेम

R V: X.121.4: These snowy mountains (arose) through his greatness (1). They call the oceans and their essence (*rasa*) as his (2). These quarters are his arms (3). Which *deva* (except him) shall we worship with offerings (4)?

R V: X.121.5: Through him heaven is forceful and the earth firm (1); He supported world of Light (*svah*) and heaven (*naka*) (2). He is the measurer (*vimana*) of the region of the midworld (3); Which *deva* (except him) shall we worship with offerings (4)?

R V: X.121.6: (Heaven-Earth) sound their thanks to him (1), for his propping them up and for his protection (2). They look upto him gratefully by their illumined minds (3),

while the Sun, rising, brightly shines over them (4).
Which *deva* (except him) shall we worship with offerings (5)?

R V: X.121.7: When the mighty Waters enveloped the universe (1),
bearing the child in birth (*garbha*) and gave birth to Agni (2),
then (Prajapati), the sole breath (*ekahasuh*) of the Gods, arose (3).
Which *deva* (except him) shall we worship with offerings (4)?

[line 3: sole breath is Prajapati]

यश्चिदापो मसिना पर्यपश्यद्
दक्षं दधाना जनयन्तीर्यज्ञम्
यो देवेष्वधि देव एक आसीत्
कस्मै देवाय हविषा विधेम

मा नो हिंसीज्जनिता यः पृथिव्या
यो वा दिव सत्यधर्मा जजान
यश्चापश्चन्द्रा बृहतीर्जजान
कस्मै देवाय हविषा विधेम

R V: X.121.8: He in his might beheld the energies (waters) (1),
bearing the discernment (*daksha*) and gave birth to yajna (2).
He was the sole God above (*adhi*) all the Gods (3).
Which *deva* (except him) shall we worship with offerings (4)?

R V: X.121.9: May he who is the father of earth protect us (1).
He created the heaven, his law of being is Truth (2).
He created the great delightful Waters (3).
Which *deva* (except him) shall we worship with offerings (4)?

[line 1: ma no himsit: may no harm come to us, i.e., may we be protected. Keith translates iy, 'may not he (Prajapati) harm us,' which is completely different.]

R V: X.121.10: O Prajapati, none other than you (1),
        has given existence to all these beings (2).
        That object of our desires for which we call you (3),
        may that be ours (4). May we become the masters of felicities (5)?

[This mantra clearly states that Prajapati is the creator who should be worshipped.]

Creation, Purusha and Prajapati
(Some ideas from two ancient texts)

Shatapatha Brahmana (6.1.1.1-5)

In the beginning, the non-existence (viz. unmanifest) alone was there. What indeed was this non-existence? The rishis verily were that. They were in the beginning as non-existent. Who then were these rishis? The vital currents or energies are the rishis. They were the ones who wore themselves out (*rishan*) by effort (*shrama*) and austerity (*tapas*). They are called as rishis because of this.

That which was in the centre (or midst) of these vital currents (viz. the chief vital current, *mukhyaprana*) is verily like Indra.

It was called Indra, because it activates (*inddhayat*) all other vital currents from there by its extraordinary power (*indriyena*, by its energy); it makes them all alive. Indra is its hidden name. Thus seven purushas came into being from the (seven) vital currents.

Then the seven vital currents deliberated: "We cannot create, with the purushas being like this (viz. multiple and disparate). Let us make one purusha out of these seven." Then they fashioned one Purusha out of the seven. From above the navel (*nabhi*), two Purushas were located at the sides (*paksha*). And the remaining Purusha was to serve as the very foundation (*pratishtha*) for the other six.

Whatever now was the wealth and glory (*shrih*) and essence (*rasa*) of each of the seven Purushas, was gathered upwards, and that became the head (*shirah*); and all the vital currents sought refuge in this part. The body is thus known as 'locus of all' (*sharira*).

This complete Purusha became the Prajapati, the creator and protector of all beings; and it is this Prajapati that is Agni, who presides over the sacrificial rites.

Taittariya Aranyaka (TA):

TA (1.23) calls the Purusha by the expression Prajapati, and describes how he became responsible for creation. At the beginning there was only water, and Prajapati took shape and floated on its surface on the leaf of a lotus plant (*pushkara-parna*). In his interior, in his mind, there then arose a desire (urge, impetus, primordial will) to create all this (what we see now). Whatever now a Person desires, he gives expression to it in speech and in action. This is the bridge between what exists not and what exists. The passage is accomplished by what is known as '*tapas*,' which word means austerity, askesis, reflection, brooding, intense heat. Creation proceeds only through '*tapas*.'

In the exhilaration of tapas, Prajapati shook his form vigorously. From the flesh-like constituents of his form came forth the sages known as Arunas, Ketus and Vatarashanas; from his nail-like parts, the sages known as Vaikhanasas; and from his hair-like parts Valakhilyas. However, the essential aspect of the watery mass solidified in the form of a tortoise and issued out. Prajapati inquired if this was the offspring of his own skin and flesh. The tortoise replied: 'No, I have been there all the time, even prior to all these beings that have now come out.' The tortoise-form was what appeared now, but the spirit of it was always there, and this was Purusha.

This indeed is the Purusha nature of the Purusha; the expression 'Purusha' signifying 'what was there earlier' (*purvamsamabhut*). The Purusha, to demonstrate his power, arose then with a thousand heads, with a thousand eyes and a thousand feet. The number 'thousand' indicates vastness and immeasurable immensity of creation.

This account appears to be an introduction to Purusha Sukta. The first words of the mantra of the sukta are repeated here, with the suggestion of the context.

अग्ने मन्युं प्रतिनुदन् परेषाम्
अदब्धो गोपाः परि पाहि नस्त्वम्
प्रत्यञ्चो यन्तु निगुतः पुनः
ते अमैषं चित्तं प्रबुधां वि नेशत्

धाता धातणां भुवनस्य यस्पतिः
देवं त्रातारम् अभिमातिषाहम्
इमं यज्ञमश्रिनोभा बृहस्पतिर्देवाः पान्तु
यजमानं न्यर्थात्

उरुव्यचा नो महिषः
शर्म यंसत्
अस्मिन् हवे पुरुहूतः पुरुक्षुः
स नः प्रजायै हर्यश्व मृळयेन्द्र
मा नो रीरिषो मा परा दाः

ये नः सपत्ना अप ते भवन्तु
इन्द्राग्निभ्यामवे बाधामहे तान्
वसवो रुद्रा आदित्या उपरिस्पृशं
मोग्रं चेत्तारम् अधिराजमक्रन्

R V: X.128.6: O Agni, driving away the wrath of foes (1),
    as guardian unfailing, do you guard us on all sides (2).
    Let your foes turn away again (3),
    and may their plotting with their foresight be ruined (4).

R V: X.128.7: May the upholder of upholders, lord of the world (1),
    the God Savita overcome the foes (of yajna) (2).

May the gods, the twins Ashvins and Brhaspati project this yajna (3),

and also (protect) the yajamana from misfortune (4).

R V: X.128.8: May the mighty one (*mahisha*), wide-extending (1), who is much invoked in this yajna (3), bring us happiness (2).
O Indra of the steeds, be gracious to our successors (4).
May no harm come to us, abandon us not (5)

R V: X.128.9: May our rivals depart (1).
With Indra and Agni we overthrow them (2).
May Vasu-s, Rudra-s, the Adiya-s make me high-reaching (3),
fierce, conscious of knowledge, (*chettaram*) and overload (4).

[uparisprsham: touching the plane above; *ugra*: fierce or violent (to the foes); All the 9 mantra-s are also in TS (4.7.14) in almost identical forms.]

Anuvaka 11: Sukta-s (129-151)

R V: .:129: Creation

Rishi: Prajapati Parameshthi

R V: X.129.1: **Waters**

R V: X.129.2: **Breathed by self-law**

R V: X.129.3: **One was born**

R V: X.129.4: **Heart**

R V: X.129.5: **Impelled by self-law**

R V: X.129.6: **Wonder about creation**

R V: X.129.7: **The creator**

[Metre: Trishtup (11, 4)]

[The Rig Veda treats the topic of creation in a very original way in several sukta-s titled as *bhavavrttam*, a crisp metaphysical thought about the beginning of creation. These hymns are (RV: X.129), (RV: X.154) and (RV: X.190) and others. In the entire Rig Veda the most famous philosophical hymn is (RV: X.129), due to the Rishi Prajapati Parmeshthi. It has seven mantra-s. The first two mantra-s refer to various view points about the beginning of creation, this sukta is not concerned with these views. The sukta does not have any criticisms. Rishi is giving his revelation. The first half of the first mantra mentions *sat* (existence), *asat* (non existence), *rajas* (principle of movement), *vyoman* (the Empyrean or ether). It asserts that it is not concerned with them. Recall that (RV: X.72.2) asserts that *sat* (beings or existence) was born of *asat* (non-existence. (RV: X.5.7) asserts:

He is the being (*sat*) and the non-being (*asat*) in the supreme abode (parame vyoman) (1), in the birth of the understanding (daksha), in the lap of the indivisible mother (aditi) (2).

What we call the ultimate or absolute, which is indicated by *tat* is beyond the concepts of existence (*sat*) or non existence (*asat*). Note that in the Hindu philosophical thinking, *sat* and *asat* are not antithetical concepts. It mentions the **One** (*ekah*) which exists by its own power (*svadha*).

Synopsis

The seven mantra-s fall into 3 groups namely mantra-s 1 and 2, mantra-s 3, 4 and 5, and the mantra-s 6 and 7. The whole sukta deals with the beginning of the creation. The question posed is, 'what is the beginning of creation?' Mantra-s 1 and 2, exclude several possibilities. The first step is not *asat* (non-existence), *sat* (existence), *rajas* (the principle of movement), *vyoman* (the supreme station). Verse 3 specifically states that the world as we see it has come out of the darkness concealed in darkness; it has come out of the deep and abysmal flood or ocean (the inconscient ocean, *apraketamsalilam*) that covered all things. Everything is hidden in this formless being owing to the fragmentation (*tuchchhyena abhu apihitam*). Note that it isUnderstood though not explicitly stated that the higher and self-luminous power descends into this ocean; it raises again out of this ocean to reconstitute in the conscient its vast unity. This One (*ekam*) mentioned in verse 3 brings to birth this world by its own greatness (*tan mihina ajayata*). In that non-existence the seers have found by desire in the heart and the thought in the mind that which builds up the existence. This non-existence (*asat*) is the first aspect to emerge from the inconscient ocean. This darkness is the Vedic night mentioned in RV (1.35.1) which holds within it all the world and her unrevealed potentialities in her obscure bosom. Above this ocean is the goal (*prayati*), below is the intrinsic power (*svadha*) which draws upward. The last 2 verses pose the question about the nature of the Supreme Being. The usual translation of the last line is, 'He knows, or he knows not.'

Such scepticism is out of place in the entire Veda. It is common knowledge in logic, that 'or' does not always mean 'exclusive or.' It can be inclusive also. Hence A. K. Coomaraswamy (A.K.C.) ('A New Approach to Vedas,' Luzac & Co, London, 1933.) translates the phrase appropriately, 'He knows and he knows not.'

RV:..:129.1: Non-existence (*asat*) then was not, nor Existence (*sat*) (1);
neither the principle of movement, nor space
there beyond (2).
What covered over all (*avarivar*) and where (3),
or what was any resting-place (*sharman*) (4)?
What were the waters (5)? Fathomless abyss (6).

[sat and asat: Note that the mantra 4 states that the existence (*sat*) was formed from the non-existence (*asat*). Recall (10.5.7).

rajas: usually it is translated as the midworld (*antariksha*). However it is the principle of movement (*gati*) caused by the chit (consciousness force). According to A.K.C. here is the earliest mention of the 3 gunas of

Sankhya namely *tamas*, *rajas* and *sattva*. *sattva* is not mentioned, but its cognates are there.

vyoma: it is *akasha* or space. Regarding the phrase '*paramevyoman*,' see (RV: X.123.5).

न मृत्युरासीदमृतं न तर्हि
न रात्र्या अन्ह आसीत् प्रकेतः
आनीत् आतं स्वधया तदेकं
तस्मात् ह अन्यत् नः परः किं चनास

The book, 'Hymn of Creation' by Vasudev S. Agarwala, published by Prithvi Prakashana, Varanasi in 1983 contains 4 translations of the Hymn RV (X:129) by eminent indologists and the commentary of Sayana. This book gives interesting excerpts from the work of Pandit Ojha entitled 'dashavada-rahasya.' The ten doctrines are outlined in the referenced book.

Sukta (RV: X.129)

avarivah: what covered over? In the Upanishad and the Veda, the world is perceived as covering the ultimate reality. Recall RV (V:62.1) (*rtena rtam aphihitam*, truth covered by truth). Also *apavrnu* in Isha Upanishad. (15).
sharman: it is related to *charma* in Shatapatha Brahmana(SB) (3.2.1.8). It means the resting place. Just as the skin covers and shelters the bones and muscles, the question is, 'what shelters the reality.' Sharman is that which offers shelter. Finally the mantra states that the first step in the creation is the waters (*ambhah*).]

RV X:129.2: Then was neither death (*mrtyu*) nor life (*amrta*) (1),
nor any sign (*praketa*) of night or day (2).
That One breathless breathed by intrinsic-power (*svadha*) (3).
None other was, nor ought there beyond (4).

[*amrta* and *mrtyu*: The two concepts are intimately related and not mutually exclusive. SB (10.5.2.4) states, '*Amrta* exists in *mrtyu* and *mrtyu* in *amrta*.' Examples of such couplets are energy, matter; divine, human; beyond time and space, conditioned by time-space; principle of rest, principle of motion [Ojha]. See also (RV:X.72) regarding Aditi and Martanda. According to SB (10.5.1.4), Surya is the sign of demarcation between *amrta* and *mrtyu*, separating the realm of immortality from the realm of death. *ahoratra*: Day and night. Here the key idea is time (*kala*). RV states that the first step in the creation is also not time. For the metaphysics of time, see AtharvaVeda (AV) (19.53.14). *ekam*: finally the mantra mentions that the chief agent is that One, *ekam*. He breathes (*anit*) without death (*avatam*). Note that the breathing involves movement. There is no concept of movement yet. Yet the life energy is there. Hence the phrase 'He breathes.' How can one do this? Is it possible? The Upanishad states that the One does it by its intrinsic power or the power of self-arranging (*svadha*). Apart from this One, there is nothing.]

तम आसीत् तमसा गूळ्हमग्रे
अप्रकेतं सलिलं सर्वमा इदम्
तुच्छ्येन आभु
अपिहितं
यदासीत् तपसः तत् महिनाजायत एकम्

RV X.129.3: Darkness hidden by darkness in the beginning was this all (1).
This all was an ocean without mental consciousness (*apraketam*) (2).
All is hidden (*apihitam*) in the formless being (*abhu*) (4),
Owing to the fragmentation of consciousness (*tuchchhyena*) (3).
Out of it, One was born by the greatness of its energy (5).

[*tama*: darkness; *gulham*: hidden; *mahi*: greatness; *ajayata*: born; This mantra describes the beginning of creation. *asat* is that which is continuously changing without any order. The reason is that the consciuousness is in fragments as it were (*tuchchhyena*). Note that in the *asat*, there is only the action of prana, but no power of mind which gives the order. Everything is hidden. Now by its own might, the supreme one is born. The first line is also in Maitrayana Upanishad.]

कामस्तदग्रे समवर्तताधि
मनसो रेतः प्रथमं यदासीत्
सतो बन्धम्
असति निरविन्दन्
हृदि प्रतीष्या कवयो मनीषा

R V: X.129.4: In the beginning, desire (*kama*) arose (*samavartat*) therein (1).
The primal seed (*retas*) of mind (*manas*), that was the first (2).
The seers of wisdom (*kavayah*) found out in non-existent (*asat*) (4),
that which builds up (*bandhum*) the existence (*sat*) (3).
In the heart they found it by purposeful impulsion (*pratishya*) and by the thought-mind (*manisha*) (5)

[The mantra states that the seers or rishi-s found in the non-existence or the inconscient ocean, that which builds (*bandhum*) the *sat*. usually *bandhu* is translated as kin. But translating it as 'to build' is much better. The idea of *sat* being born of *asat* is elsewhere in the Veda. They found that power in the heart as the impulsion (*isha*) and in the mind as the thought (*manisha*). The Vedic triplet (*hrda, manasa, manisha*) occurs in RV (1.61.2) and Katha Upanishad. (2.3.9)]

तिरश्चीनो विततो रश्मिः
एषाम् अधः स्विदासीत्
उदुपरिस्विदासीत्
रेतोधा आसन् महिमान आसन्
स्वधा अवस्तात् प्रयतिः परस्तात्

RV 129.5: Their ray (*rashmi*) was extended horizontally (1).
There was something above (3),

there was something below (2).
Seed (*retas*) was, all-might (*mahimanah*) was (4);
intrinsic-power (*svadha*) below, purpose (*prayati*) above (5).

[*svadha*: intrinsic power; power of self-arranging; Note that the creation is impelled by the intrinsic power from below; the goal of creation (*prayati*) in the station above pulls up the consciousness levels to manifest and establish the truth everywhere.]

को अद्धा वेद
क इह प्र वोचत्
कुत आजाता कुत इयं विसृष्टिः
अर्वाग्देवा अस्य विसर्जनेनोद्धा
को वेद यत आबभूव

RV X:129.6: Who knows it aright (1)? Who can here set it forth (2)?
Whence was it born (*ajata*), whence poured forth (*visrshtih*) (3).
These gods (*devah*) are from its pouring-forth (*visarjana*) (4),
Whence then it came-to-be (*ababhuva*), who knows (5)?

[This mantra and the next are viewed by some translators as indicating scepticism since they pose the question beginning with who. As A.K.C. points out, these questions indicate only wonder. This creation is so wonderous that we cannot even think about the One (or Supreme) who made it possible.]

इयं विसृष्टिर्यत आबभूव
यदि वा दधे यदि वा न
यो अस्याध्यक्षः परमे व्योमन्
त्सो अङ्ग वेद यदि वा न वेद

RV X:129.7: From what source did this creation (or discharge) (*visrshtih*) came into being (1)?
or whether one appointed (*dadhe*) it or not (2).

> He who is over-eye thereof in Supreme station (3),
> he knows indeed, or knows not (in advance) (4).

[The parts (3) and (4) are very interesting. The usual translation done by Indologists is 'he knows indeed or he knows not.' They are happy to note that even the creator does not know all. The sole exception is A.K.C. He translates it 'He knows and he knows not.' The idea is that in every act, the outcome is not really fixed at all till the last second. The grace can act at the last minute. there is no limitation.

adhyaksha: over-eye over-seer; *paramevyoman*: the infinity of the superconscient being; 'Ether' in old translations. See (10.123.5).

We have heard of the adage that 'not a blade of grass moves without His consent.' It is true. But this statement does not state that everything is planned in advance. In every action, there are so many possibilities for its termination. Only the Creator decides which possibility will prevail. The Creator does not need to plan ahead. Thus both the statements 'he knows' and 'he knows not in advance' are true. He does not specify the way of conclusion of an action in advance, since such a specification limits his Own Power. See X:131.3 movement of truth is not fixed in advance in detail). By definition, the Supreme Person has no limitations.]

CHAPTER 6

THE CREATION IN RIG VEDA X:129

by Steffen Stenudd

THE PARADOX OF ORIGIN

Few cultures are as impenetrably complex as that of India. This is evident also in its ancient sources to ideas of the creation of the world in Rig Veda, the collection of hymns from around 1500 to 800 BCE, the poet of one of them contemplates the very question if something can be first, i.e., if there can have been a creation at all.

Rig Veda 10:129 is in a famous hymn of the tenth mandala. It is generally regarded as one of the later hymns, probably composed in the 9th century BCE. It has the Indian name *NasadiyaSukta*, "Not the Non-existen", and is often given the English title *Creation*, because of its subject.

The Paradox of Origin

The advanced abstract reasoning in the hymn has brought it a lot of attention, not only within Indology, but from scholars of philosophy and the history of religion as well. Its line of thought relates splendidly to cosmological thinking of the philosophers of Ancient Greece, all through to present day astronomy.

And it ends with what seems like a punch line, a paradox taken to the extreme, almost as if the unknown poet of it was making a joke. Here are the last lines of it (in Max Muller's translation):

> Who knows from whence this great creation sprang?
> He from whom all this great creation came.
> Whether his will created or was mute,
> The Most High seer that is in highest heaven,
> He knows it – or perhapce even He knows not.

Mainly Rig Veda X:129 reveals an insoluble paradox in which the human mind of the past as well as the present easily gets trapped: How can the universe have sprung into existence, i.e., how can something come out of nothing? How can there be a beginning, before which there was nothing?

Much of what puzzled people three thousand years ago, still puzzles us today. This is dilemma, too. Present-day scientists wrestle with the paradox, speculating about multiverses and such in an effort to explain the something out of nothing. Doing so, they might just move the problem to another location, not solving it at all.

The Creation in Rig Veda 10:129

A Synthesized Version of the Hymn

After examining seven major translations of Rig Veda 10:129, I need to compose a synthesized version of the hymn based on them, before moving on to making suggestions about how it might be interpreted. So, here is the hymn as I perceive it.

As implied by my comments on the seven English versions presented on this website, parts of the hymn are uncontroversial with very similar interpretations from all the seven translators, whereas other parts are much more complicated. A pattern emerges, but I return to that later.

Below is my synthesized sketch of the hymn, where I have tried to give it as much clarity as possible. Mainly but not always, I have followed the majority of the seven translators where that could be calculated, but also I have allowed myself some freedom where I felt it was called for in seeking out what the poet of the hymn is most likely to have intended.

Of course, since I am not an Indologist my choices are based on other things than any profound knowledge of the original poet's context. Apart from the arguments of the translators, I have mostly considered the hymn's inner logic and my understanding of patterns in creation myths as well as ancient cosmogonic thought.

I will discuss my choices below. Here is my synthesized version of the Rig Veda 10:129 creation hymn:

Rig Veda, Mandala 10, hymn CXXIX. Creation.

Nasadiya Sukta ("Not the non-existent")

1

There was neither existence nor non-existence then.
There was neither sky nor heaven beyond it.
What covered it and where? What sheltered?
Was there an abyss of water?

2

There was neither death nor immortality.
There was nothing telling night from day.
The One breathed breatless autonomously.
There was nothing else.

3

There was darkness concealed in darkness.
All was water without shape.
The One enclosed in nothing
Emerged by the power of heat.

4

First to arise was desire,
The primal seed of mind.
Wise poets searching their hearts
Found the bond between existence and non-existence.

5

That cord was stretched across.
What was above and what below?
Seeds were shed and mighty powers rose.
Below was urge, above was will.

6

Who knows and who can here tell
Whence it all came, whence is this creation?
The gods came later to this world.
So who knows whence it came?

7

Whence this creation came,
Whether he made it or not.
The overseer of it in the highest heaven,
Only he knows it. Or doesn't he know?

First Verse Choices

As for the first verse, its content is not that controversial. The seven translators differ only in minor details, not in the general statement of the verse. Mainly, they describe the yet non-existent cosmos in different terms.

They all make it clear that both references in the second line are to celestial components. They just name them differently. The hymn suggests that there are two levels of the celestial yet to be created. I choose the words sky and heaven, in the hope of giving the impression of one layer closer to earth than the other. That's what the hymn specifies quite clearly, whatever English words we use.

Nothing else in this verse causes any great conflict between translations.

Second Verse Choices

Also regarding the second verse, the seven translators pretty much agree. Their choice of words may vary, but not really so that they alter the meaning of the verse. Therefore, I had no trouble finding a straightforward way of writing it, which I trust that none of the seven translators would disapprove, at least not with any emphasis.

Everyone except Joel P. Brereton chooses immortality as the opposite to death. He calls it deathlessness, which is interesting in its much more limited assumption, but I go with the majority. Furthermore, it is very likely that the poet expected the audience to think about mortal earthly beings on one side and immortal divine creatures in heaven on the other.

As for night and day in the next line, A. L. Basham has explicitly mentioned the light (in the form of a torch), since that is what makes day emerge. The others haven't, but talk about the lack of a sign or indication. Anything specific would assume more of the cosmology than what seems appropriate at this time in the hymn.

There was no sign of either night or day, but the hymn doesn't seem to venture into what that sign would be, specifically. So, I don't specify it, but the text of the hymn indicates that this obscurity is in the perspective of a potential (but non-existent) observer. So I implied it.

The line where the One is introduced causes some problems for the translators, but they mainly agree on what is stated also in my synthesized version. The One breathed without air, since that was yet to be created, and did so without any outer means.

I know the term "autonomously" lacks the ancient flavour, but I chose it to avoid specifying gender. Although the One is later described as male, it doesn't really apply to the chaotic state before creation commences and the One is awakened to become self-aware.

Third Verse Choices

In the third verse, the seven translators agree pretty much about the first two lines, but not so with the following two. That's when creation commences, which is a vastly important moment – especially in this poem, with its punch line of that being hidden to all.

It's not even so that all translators specify the role of the One in this, though it still might be implied. Max Müller talks about a germ, H. H. Wilson about a united world and Ralph T. H. Griffith a unit. Still, it's safe to say that they all refer to the mysterious One mentioned already in the second verse of the hymn.

The translations divide as to what makes creation begin. Most of them say that heat got the process going, but Wilson suggests austerity. Griffith says warmth, which is just heat with a milder choice of word. As for how they form this sentence, none is identical to the other.

In spite of these differences, what is described is quite the same – with the possible exception of Wilson. Creation begins with a rise of temperature. That may be symbolical, but the text hints nothing of it.

My problem was with the word "emerged." I considered an awakening, since that seems to be what the One goes through, but it would presuppose too much. The seven translators have different solutions: "burst forth," "produced," "born," "came into being" and "arose." I did not want to use the concept of birth, since the One is said to already exist in some way or other – even breathing. It is an awakening, but without any support from any one of the translators, I had to settle for emerge.

Fourth Verse Choices

Desire as the primary force of creation is uncontroversial. Only Müller chooses another word, love, which is probably just in an effort not to challenge the prude ideals of his time. As for the seed in the second line, it's the choice of five translators. Joel P. Brereton uses "semen" instead, which is probably quite correct. Again, the other translators may have wanted to soften the sexual innuendo.

I stick with seed, not only because of the majority of translators, but to avoid gender specifics at this stage. The verse describes desire as the first seed of mind. The metaphor of a seed of mind makes more sense as a starting signal to the growth of a whole world, than semen of the mind would.

The latter would suggest self-fertilization, like that of Chepre in an Egyptian creation myth. But the One of Rig Veda 10:129 is a different creator, triggered by desire to move from a vague passive existence into action. The One's creation is more of an outward activity than Chepre's inward fertilization.

As for the poets of the next line, four of the translators chose "sages" instead. Joel P. Brereton points out that the Sanskrit word indicates not just any poets, but those with profound understanding. So, "wise poets" is a way of combining those traits.

It is interesting that these poets searched their hearts. That might be a Western interpretation, since our tradition gives the heart a role rightly belonging to the brain. We have seen it as the seat of moral, courage, compassion, and so on. We still use the heart as a symbolic home of such things.

That might have influenced the translators in their choice, but all seven do the same, so it must be well-founded. Also, the Western tradition is far from the only one giving the heart credit for things of the brain. That is probably due to its tendency to change the frequency of its beats when we are overcome by some sentiments.

The last line of the fourth verse has created some diversity among the translators. Most of them describe a link between what is and what is not, albeit with slightly differing wordings. But three of them – Macdonell, Doniger and Brereton – describe existence as if it was contained within non-existence. That could be implied by "darkness concealed in darkness" of the third verse, existence as a hidden potential inside non-existence, but I find it unlikely.

The dynamics of creation described by the hymn is a process where opposites have to appear and take shape, since one cannot exist without the other. Neither existence nor non-existence is possible without its counterpart. That would be the bond between them, and that makes the world take form, as stated in the first line of the fifth verse.

Fifth Verse Choices

All translators except Müller choose the expression "their cord" in the first line of this verse. That makes it unclear if the poets or the two opposites of existence and non-existence are intended. I have no doubt that the latter is what the poet intended. To avoid misunderstanding, I write "that cord."

The bond between existence and non-existence is a cord extending through an emerging cosmos, as the opposites manifest themselves by the necessity of their interdependent opposition. Surely, that's a division into heaven and earth, which is what almost every creation myth describes.

So, what was above and what below was no real mystery to the original audience of this hymn. But the question is not just rhetorical. It points out the many remaining mysteries regarding the heavenly as well as the earthly, also in particular the uncertainty of their roles in the continued process of creation. It's no wonder that the translations of this verse differ quite a lot.

They are all rather clear about some kind of fertilization taking place between the above and the below, where the former must surely be the seed and the latter the womb. But in describing the charateristics of those two, the translations diverge-even contradicting each other. The hymn describes both as powers, not necessarily one superior to the other. I call the upper one "will", since the hymn seems to see it as an initiator. It is, after all, the seed. I call the lower one "urge", because several of the translators indicate that the system allows it some say in the process, not just being a passive receiver. It engulfs the seed from above and uses it in the following procreation. The below has an urge to take part in creation.

Sixth Verse Choices

The complications of the fifth verse are gone when the sixth is reached. There is wide consensus among the translators as to how to word it and what it means. The poet suddenly interrupts the narration of creation and returns to the question asked in the beginning: who can really say what happened?

The word "here" points out an audience to the poet in a new setting. Five of the seven translators use it. So do I, since it points out so clearly that the poet brings the audience back to the present, far away from the distant moment of creation. Here and now, instead of back then, who can say what happened? The shift of perspective makes the question even more mind-blowing.

The use of "this creation" in the next line has a similar function. It points to the contemporary observers if the world they live in.

My problem with the verse is a minor one – the word "creation." All the seven translators use it. Actually, six of them do it twice. What makes me hesitate to use this word is all that it implies, such as an active and intentional process of making the world, which in turn also suggests a maker. We should not jump to the conclusion that all this is what the poet had in mind.

I checked the Sanskrit word (at spokensanskrit.de) and there is no doubt that "creation" is a proper translation of it. But the Sanskrit word (*visrsti* or *visarashti*) also means: offspring, emission, discharge letting go, quitting, leaving, allowing to flow, production, and giving. So, it seems to have other connotations then the divine magic of making the world appear, as we tend to see it in the Judeo-Christian tradition.

The Sanskrit word suggests a creation process similar to feeding: once the world is fed it pretty much takes care of its own continued evolvement and multiply. As implied in the fifth verse, the creation in Rig Veda 10:129 is one where both of the above and below are active. Maybe the latter soon takes over the process, whereas the above is reduced to a passive source of energy for it. That's quite different from the Genesis I and II descriptions of creation, where god is the only one making things happen for quite a while, until every creature on earth has been produced.

As for the second use of the word "creation" in this verse, the translators vary. Müller puts it in the fourth line and five others in the third. Griffith uses synonyms. That gave me the freedom to avoid it completely. The third line simply states that the gods were introduced later, when there was already a world or cosmos of sorts. The fourth line refers back to the second line and it makes sense that the wording is at least similar. This is the choice of most translators.

It is also the way the seventh and the last verse starts.

Seventh Verse Choices

There is even less controversy among the translators about the last verse of Rig Veda 10:129 than there is about the previous one. They all come quite close to one another in the wording and definitely so on what is implied.

It starts with the same enigma by which the sixth verse ended. Where did it all come from? As for the overseer in the highest heaven, the hymn doesn't state so firmly but it makes sense that it is the One of previous verses.

Then it is confusing that the One as a creator is put to question. Wasn't that already determined by the third verse? Actually not, the One is the first to emerge, but the following verses don't specify being's active role in creation. It seems almost to happen by itself, at least to begin with.

So, this question in the last verse is not completely out of the blue.

What the hymn ends by concluding is that not even the very first being to appear is sure to know how it happened. This punch line, certainly the main reason for the poet to write this hymn, has to end with a question mark. Only one of the seven translators, Joel P. Brereton, does so.

He even uses a triple-dot ellipsis to stress the paradox more, which is fine by me. The only reason I don't do the same is that it is such a modern technique. It becomes odd as the last sentence of a 3,000 years old hymn from India. But I bet that the poet had a similar effect in mind when writing the last line-prabably giggling when doing so.

Conclusions about the Hymn

Some Conclusions about Rig Veda X:129

The Rig Veda 10:129 creation hymn may very well be the most famous of all the Rig Veda hymns, at least internationally. It is regarded as obscure although its punch line is quite obvious. Here are some conclusions I make after studying the aforementioned seven English versions of it.

The hymn starts by stating that not even nothing could have existed before the world emerged. Therefore it can't come as a surprise that the hymn ends by suggesting that no one can know how the world was created.

No one born out of the creation can know what that primordial state was like. According to the hymn, it's not even sure that its maker-if there is one-can know it. That's almost saying there wasn't a maker, since a conscious creator must have known.

Like Big Bang

This would be very much in line with our present understanding of the birth of our universe in the Big Bang, where also the time we experience started. So there was no before, really, and therefore nothing to be known about it. And certainly, the Big Bang theory leaves no room for a conscious creator.

But the Rig Veda X:129 hymn was composed some 3,000 years ago. At that time, in India as well as anywhere else, the idea of a creation without a creator would be very radical, if not absurd.

Well, that's what we surmise about our ancestors. Upon close examination, it's not only this Rig Veda hymn of that distant past seeming to consider the idea of a creation without a conscious and active creator.

Already primordial man, back at the dawn of our species as we know it, must have found many things in nature happening without any sign of someone making it so. People back then would not have regarded natural events as some sort of machinery, before we were able to invent our own. But a world where much happened without any willful interference must have been possible for them to conceive. That might even have been what they regarded as the normal state of things.

It is not at all unthinkable that the personification of natural forces into some hidden deities, of which mythologies are full, was introduced later in human history. If so, it was probably as a way of trying to deal with the unpredictable strikes of fate. The belief in deities led to worship of them, which was to quite an extent an effort of pleasing them and thereby fate.

Be that as it may, and we can only guess about it, but Rig Veda 10:129 can certainly be interpreted as suggesting a creation by impersonal forces. It even says that creation was inevitable, because of the primordial state being impossible.

That's announced already in the first line of the hymn. There was neither existence nor non-existence, which is impossible. So, both had to appear. Something had to exist in order for nothing to exist as well, and vice versa. A state with none of them just could not be. The world burst out of necessity.

How Can the Poet Know?

So, the first statement about the impossible primordial state fits nicely with the ending statement about nobody knowing of the world's origin. It's the stuff in between that complicates things, for the simple reason that it describes what has been declared indescribable.

The hymn says that not even the gods can know how the world emerged, since they were not there at the very beginning. Not even the overseer in the highest heaven – the god of gods, so to speak – is sure to know. So how can the poet?

Why does the hymn spend several verses describing the process leading to the creation of the world, when it has also stated that this cannot be known by anyone – certainly not a mere mortal? This contradiction within the hymn is as grave as can be, since it strikes at both its outset and its final punch line. Would any poet make such a blunder?

I see two plausible explanations: Either we misinterpret the content of the hymn completely, or it has been altered since its original writing.

Misunderstanding

Starting with the first possibility, can we misunderstand the poet's claims so fundamentally – and if so, what part of it do we get wrong? Is he not at all claiming that the way the world emerged is beyond anyone's knowledge, or is his description of that emergence not what it seems to be?

The former alternative is not that likely at all, since the whole point of the hymn is its punch line statement about nobody knowing how the world emerged. Without it, the punch line loses its punch and the hymn its meaning.

Furthermore, we have no problem seeing and appreciating the paradox of an uncreated world neither being nor not being. It's not a primordial state easy to imagine, but it's very easy to see how it must be beyond anyone's understanding. Both the poet and his audience would enjoy it and regard it as the hymn's major quality.

So, could the hymn's description of how the world emerged be something else than that? A simple explanation would be that it starts when the world creation is well on the way and at least the One has awakened and can witness it. But that's sort of a post-creation, another time than the moment of which the hymn makes its paradoxical statements. Why would the poet make conclusions about one event and still spend verses describing another event?

Also, the text of the hymn firmly suggests that the description is indeed that of the very beginning of creation, before which there was only an impossible chaos.

Primordial Mystery

The only thing that would make some kind of sense is if the hymn doesn't claim that creation is incomprehensible, but the primordial state before it began. That is the situation described by the first lines of the hymn, until the One is introduced in the second half of the second verse. Or, if we accept the idea of the One belonging to the primordial state, all the way until the last line of the third verse, where heat makes the One emerge.

That's a possible solution to the problem. The poet claims that we know how the world was created, but not out of what – not what was before it.

That's sort of the same problem as we have with the Big Bang theory, unable to describe what preceded the Big Bang and where that might have been – or not have been. A modern astrophysicist would totally get that message of the hymn and agree with it. It also touches the problem the philosophers of Ancient Greece discussed quite a lot: if something can come out of nothing.

It makes perfect sense if the poet claims that nothing can be known about the world before its emergence. As soon as the world started to emerge, though, its process would be traceable – at least in theory.

Modern astronomy has a similar limit at 10^{-43} seconds into Big Bang, before which it is very hard or impossible to speculate about its properties (the Planck epoch). After that very short period of time, the universe started to behave in a fashion orderly enough for calculations to describe it well.

The punch line of the Rig Veda hymn might be about what was before creation, not during it. Would that be a possible interpretation of the Sanskrit text?

There is nothing contradicting it in the first five verses. As for the remaining two, it is a bit unclear but the translations actually open for this interpretation.

In the sixth verse, only one of the seven translators, A. L. Basham, writes "how creation happened." All the others use the word "whence," from where, which does point to the primordial state being the mystery, and not the actual process of creation. So, six out of seven versions indeed point to this understanding of the hymn.

The seventh verse refers back to the sixth verse as to what the question is. Therefore, it conforms just as well to this interpretation.

So, indeed, it is quite possible to interpret Rig Veda 10:129 as stating that nothing can be known about the universe before its creation, whereas a lot can be concluded about how the creation played out.

Then the content of the hymn can be described like this: What was before the world emerged is impossible to know. Here is how the world emerged from that mystery – but again, what was before is impossible to know.

We can all agree to that.

Later Additions

Another possibility is that the hymn has been edited since its original writing, by others than the original poet. That would easily enter contradictions. Some aspects of the hymn indicate it.

Already when I started studying the Rig Veda 10:129 hymn, I was wondering about the clarity of its beginning and end, versus the complications in between. As mentioned above, the beginning and the end conform to the thesis that nothing can be said about the creation of the world. Still, the middle verses outline that creation, as if the poet knows it well.

I have also noticed in the seven English translations treated on this website that they mainly agree on the beginning and end of the hymn, but vary much more regarding the middle, almost as if wrestling with another text. It seems that the Sanskrit original has a clarity in the first and last verses that is obscured by the other ones. That does suggest the possibility of some tangling with the hymn, after its original composition.

Furthermore, there are some obvious contradictions appearing in the middle of the hymn. The most blatant one is about the primordial abyss of water. The first verse questions if there was such a thing, but the third verse states firmly that it was there.

The presence of the One is a similar anomaly. In the total primordial nothing, can there really be a One, who is even breathing albeit breathlessly? The One is introduced in the second half of the second verse, as a creator to be. But except for being awakened by heat and feeling desire, the One plays no obvious role in the creation of the world. That seems to happen out of its own necessity, whether desired or not.

The One – or anyone else – as a world creator is questioned in the seventh verse, and the sixth says categorically that none of the gods existed at the time when the world started to emerge.

It should also be noted that the overseer in the highest heaven, mentioned in the seventh verse, is not necessarily the same as the One. Probably not, or the same term would be used for that being.

There is also an odd shortness to the description of the world creation, as if intended to be more elaborate at the outset, but then stopping short and returning to the statement that nothing can be said about it.

Indeed, the middle of the hymn has several inconsistencies compared to the beginning and the end.

Edited to Conform

It would not be too hard to explain why such editing took place, if that is the case. The hymn's beginning and end almost give an atheistic sense in their absence of gods. This would provoke anyone confessing to a world run by deities. That was surely the case with most people of ancient India, at the time of this hymn's writing. Maybe even more so, later on.

When there is a pantheon and the worship of it, there are strong forces in that culture working to conform it to that belief. Hymns of the Rig Veda would be targets of this, certainly a hymn ignoring the role of the gods in the very creation of the world.

Joel P. Brereton gives an example of addition to Rig Veda 10:129 in his text on this hymn, presented on this website. There is a Sanskrit version where two verses have been added after the seventh, elaborating on specific divine activity in the creation of the world. So, what's to say it has not happened to the seven verses we regard as the original Rig Veda 10:129?

The extant copies of the text are far younger than the estimated period of its composition. Many things can have happened to it through the many centuries between its writing and our oldest copy of it.

Actually, there could very well have been some editing of it already when it was entered into the Rig Veda collection, so that it would conform to the other hymns as well as to the cosmological beliefs of those putting together the collection.

The Hymn Stripped Bare

If we edit out any description of the process of creation, including the role of the One, what remains of the Rig Veda X:129 hymn is this (in my synthesized version):

> There was neither existence nor non-existence then.
> There was neither sky nor heaven beyond it.
> What covered it and where? What sheltered?
> Was there an abyss of water?
> Who knows and who can here tell
> Whence it all came, whence is this creation?
> Whether he made it or not,
> The overseer of it in the highest heaven,
> Only he knows it. Or doesn't he know?

The third, fourth and fifth verses, describing the emergence of the world, are removed. So is the second verse, because it made statements about the primordial state, as opposed to the open questions of the first verse. That seemed both out of place and repetitious.

I also removed the mentioning of gods in the sixth verse as well as the repetition of the question of its first line in the fourth. That made the first sentence of the seventh verse redundant, too.

So, what is left is the complete first verse, half of the sixth and all but the first line of the seventh verse.

I don't know what this does to the meter and poetic form of the hymn in Sanskrit, but this sharply shortened version makes a lot of sense in English. Its beginning explains the paradox of the primordial state, which makes it impossible to describe it – or even in what way it would lead to the emergence of the world. And that leads directly to the punch line ending of the hymn.

As far as I can see, this or something like it could very well be the original content of the hymn, before pious scholars felt the need of adding a cosmogony fitting their own conviction.

Both Alternatives Plausible

So, which one of the above alternatives is the likely explanation to the inconsistencies of the Rig Veda X:129 hymn? Should its message be understood as only the primordial state being beyond understanding even to the mightiest of deities, or has its inconsistencies entered with verses added by others than the original poet in their effort to make it conform to their thoughts?

It could be both.

Since only one of the seven examined English translations disproves the former alternative, there is good reason to accept it. The hymn as we know it today points out that it is not creation that is hidden in impenetrable mystery, but whatever preceded it. Not how creation happened, but out of what.

That would be enough to explain the hymn if it weren't for the remaining contradictions I have mentioned above. Those contradictions are to be found in the part of the text describing the process of creation. This theme is awkward in a hymn with the punch line of Rig Veda X:129, so it does seem out of place. Why bother with a description of the creation process in a hymn about the complete enigma of whence creation came? And when doing so – why the contradictions?

So, it is quite plausible that the hymn has been altered not only once and not only by one editor. Either that or a sloppy poet did a poor job with the first and only alteration.

Two Creation Stories

There does seem to be two partly contradicting narrations of the creation colliding in this hymn. Furthermore, those two fragments of creation fit two separate types of creation myths that have many examples around the world.

The first fragment of a creation story is that about the One, who is awakened by heat and immediately feels desire. The desire for company, surely, and the desire to change the original chaos into a decent habitat. That motivation in a primordial creator deity is found in many myths.

A good example of this kind of creation is the Xingu myth, also presented on this website. In the beginning, according to the Xingu Indians, there was only Mavutsinim, who was all alone. He turned a shell into a woman and mated with her. But when she bore him a son, he left with the boy. The woman cried and turned back into a shell.

Something similar is to be found in Genesis II of the Bible, where god creates Adam and Eve and all the animals, filling the Garden of Eden – not unlike a child playing with a dollhouse.

Both creation myths suggest a world creator acting to get out of his loneliness. The One of Rig Veda X:129 is similar, having desire immediately on becoming aware. In this mood, the One would surely hurry to create company.

But nothing is mentioned of that. Instead, the Rig Veda hymn shifts to another perspective, where the One is absent.

An Impersonal Creation

The second fragment of a creation is quite different in nature. It is impersonal, with the bond between existence and non-existence causing a cord to divide the primordial chaos into an above and a below. This is rather abstract, but again not that far from the cosmogony of several other creation myths. Ancient cosmological myths are far from only filled with spectacular deities acting from very human incentives.

It also has to be remembered that many deities in old mythologies are representatives of natural forces. So, even a creation story full of active deities can be seen as forces of nature causing this and that to happen, without any initial conscious will to dictate it.

This possibility must be considered also regarding the One of Rig Veda X:129. The mysterious primordial being, described so sparsely, could very well be the personification of a natural force, or a certain characteristic the primordial state must have in order for the world to emerge from it. Aristotle would say that the primordial state would need to contain a cause, or there would be no effect of it. The One is hinted as the primal cause leading to creation, to be compared to Aristotle's unmoved mover.

This ancient way of reasoning deserves another example, which is that of the sun. Its effect on earthly life is obvious and abundant. Mythologies turning it into a divine creature with its own eventful past must have come much later than mankind's awareness of the sun and the effects of its light. Also the worship of it must have been something evolving by time, maybe at first as little more than what we do when we greet the morning and hope for sunshine.

Most deities in most if not all mythologies are linked to natural forces, celestial bodies and so on. They are symbols of these natural phenomena. Often in the myths their adventures represent natural events in an effort to explain them, or maybe just to play with them.

Comparing again to the Bible, Genesis II explains how the serpent lost its arms and legs, so it had to crawl forever on its belly.

Mythologies are full of such explanations, and so are folk tales of much later origin. We should not take for granted that these stories already at their invention were firmly held beliefs. Many of them could simply be playful figments of the imagination, pure entertainment not meant to be taken seriously. Nocturnal storytelling by the fire.

Ancient Abstractions

Of course, there was also a search for explanations of the world and all its phenomena. Where did it come from and why did things behave as they did? These questions often found answers proving advanced abstract thinking, such as that of the world emergence in Rig Veda X:129.

I would say that abstract conceptions of the world creation are likely to be older than those involving deities and other characters, simply because the latter indicate further developments of the story. But there is no consensus about it among researchers of myth. A widespread assumption for the last two hundred years of the study of myths has been that the more personal the forces at work are, the older the myth is. I seriously doubt that.

As for Rig Veda X:129, I would be inclined to regard the more abstract second fragment of creation to be the older one, and the short part about the One being of later invention. Whatever the case, let's have a look at what the abstract version has to say.

Creation and Procreation

The Rig Veda hymn speaks of a bond between the existent and the non-existent, a cord stretched out dividing the world into an above and a below. Seeds were shed, most likely from above, and thereby mighty powers arose below. The dynamics between the above and the below is described as will in the former and urge in the latter, though the translators have used widely different expressions.

This short sketch of creation is quite recognizable. It compares well to many other creation myths. That is what should be expected, for the very simple reason that it is mainly the same world observed and the same ignorance of the scientific facts we have piled up during the last few centuries. Ancient people speculating about something as distant as the creation of the world were not thinking that very differently, at least not beneath the complexity of developed mythologies.

The division into an above and a below is found just about everywhere. It is also something any human being could easily observe. Heaven and earth. The former nourishing the latter is also familiar and explained already by the obvious influence of the sun and the rain. Ancient man was well aware of earthly dependence on heavenly forces.

But anyone could also see that earth contained powers of its own. There was the mighty sea, there was procreation continuously going on among plants and animals. The spark of existence may have come from the above, but its continued process was something mainly handled by all the creatures of the below. It's the difference between creation and procreation. So, the initiative was from above, but the urge to continue was below.

What little is described by this fragment of creation in Rig Veda X:129 would have made much sense to ancient man also outside India. What it describes is just the very start of creation. This short hymn would lose its character if it elaborated on what happened next. Also, its punch line would soon become irrelevant.

Either the original poet of the hymn realized this and stopped short, or a later editor of the text settled for just adding the minimum of a creation process deemed necessary. My guess is the latter.

Three Poets

So, to sum it up: I suspect that Rig Veda 10:129 is not a hymn composed in its entirety by one poet, but the work of three. The initial text would be the hymn's beginning and end, declaring that the primordial state before creation was so absurd that it cannot be understood by anyone – not even by the creator of the world, if there was one. Another poet added the lines about the One, implied to be the creator of the world, probably to conform the hymn to the cosmological belief of that poet. A third poet had an alternative view on creation, seeing it as an impersonal process out of necessity, and made that addition to the hymn.

The third poet's addition could very well have been made before the lines about the One. The abstract thinking of that addition is much more in line with the original poet's perspective – so much that I would even consider the possibility that those lines were also written by the original poet.

What sticks out the most is the part about the One. That implies a primordial and superior deity, which would explain why a later editor felt the need to include it. It would be an editor worshipping that deity.

We have many examples of such revisions of old myths. It happened in Enuma Elish, presented on this website, where the Babylonian god Marduk was replacing older ones as the highest deity and also given a prominent role in the creation. Something similar was done with Yahwe in the Bible, doing away with the pantheon of older gods in the Middle East.

India, with its great complexity of deities and myths as well as its long history of thought, must have several examples of the same. Rig Veda 10:129 is likely to be one of them.

The Creation in Rig Veda X:129

Various authors who translated Rig Veda:

1. **The Creation in Rig Veda X:129**
2. **Max Müller's Translation of Rig Veda X:129**
3. **H. H. Wilson's Translation of Rig Veda X:129**
4. **R. T. H. Griffith's Translation of Rig Veda X:129**
5. **A. A. Macdonell's Translation of Rig Veda X:129**
6. **A. L. Basham's Translation of Rig Veda X:129**
7. **Wendy Doniger O'Flaherty's Translation of Rig Veda X:129**
8. **Joel P. Brereton's Translation of Rig Veda X:129**
9. **The English Versions of Rig Veda X:129**
10. **Table of Seven English Versions of Rig Veda X:129**
11. **A Synthesized Version of Rig Veda X:129**
12. **Conclusions about Rig Veda X:129**

CHAPTER 7

THE CREATION OF THE UNIVERSE
– THE BRIHADARANYAKA UPANISHAD

by Sri Pujya Krishnananda

CHAPTER I

Second Brahmana: The Creation of the Universe

Now follow some very difficult symbols of the Upanishad. Literally, they cannot be easily grasped. Even the Sanskrit is not classical; it is highly archaic. It is a vedic language. And the idea conveyed through this most difficult style is still more difficult, so that one cannot easily make out the sense of some passages, unless we deeply think over the words as well as the meanings that are hidden between the lines. An unphilosophical mind may not be able to understand the hidden meaning of these symbols, and perhaps it is the case with all symbols; they cannot be understood literally.

The symbolic description here is one of the processes of creation. How things come; and what it is that we see with our eyes. Where are we living? What is the connection between the effect and the cause? What is our connection with the Universal Being? What is the relationship between the individual and the Absolute? All these points are discussed in a pithy and pointed way, in a few passages, commencing from the Brāhmana, or the section of the Upanishad that we are to study now.

1. naiveha kimcanāgra āsít, mrtyunaivedam āvrtam āsít, ašanāyayā, ašanāyā hi mrtyuh; tan mano'kuruta ātmanví syām iti. so´rcann acarat, tasyārcata. āpojāyanta, arcate vai me kam abhud iti; tad evārkasya arkatvam; kam ha vā asmai bhavati, ya evam etad arkasya arkatvam veda.

Originally, there was nothing. Death was enveloping everything. That is all the meaning, literally, of this sentence. In the beginning of things, what was there? Nothing was there. There was a devouring, all-consuming death principle, as it were; nothing else can we conceive. In the Veda, also, there is this very same point reflected in the Nāsadíya Súkta, which proclaims that, in the beginning, there was neither existence, nor non-existence. What was there, originally? Darkness enveloped, as it were, because there was not the light of sensory perception. What we call light is nothing but the capacity of the senses to perceive. When the senses cannot perceive, we say there is no light. In pitch darkness, a kind of light exists; but the eyes are incapable of catching the ray of that light. That frequency is quite different from the one that is necessary for the eyes to perceive. So, when there was no possibility of external consciousness, when there was no sensory activity, when there was no distinction between the subject and the object, when the seer was not distinguishable from the seen, what was there? We can imagine for ourselves, what can be there. If we are not to perceive anything outside, what would be our condition? We cannot imagine it, because such a condition has never been seen; but it would be a veritable abolition and obliteration of all consciousness, because every kind of consciousness is equivalent, in our case, with externally. Therefore, in the condition of non-objectivity which is the origin of things, the cosmic beginning of things, where the distinction between the seer and the seen was not marked, where the one commingled with the other, where one entered the other, where the two could not be distinguished, for reasons obvious, what was there? Nothing was there! ***Naiveha kimcanāgra āsīt***: Originally, nothing was there, because our idea of 'something' is an 'object.' There is no object present, because the object enters the subject, and vice versa. What was there, then? If nothing was there, could you tell me that it is capable of definition in some way?

The devouring death principle is the element of hunger which grasps objects. Here, hunger does not mean merely the appetite for edible dishes like rice, barley, etc. Here is a metaphysical principle. Here, the hunger is a cosmic element. It is not an operation of the biological spleen or the liver or the stomach of the individual. What is here intended is the principle of grasping. The object can be regarded as the hunger of the soul of the individual. There was nothing except the desire to grasp the object, if at all one could say that anything was there. Ašanāyayā is the hunger of the individual to grasp, absorb, contact, abolish and devour the object.

Now, this is a condition which cannot be easily analysed, unless we pause for a while on this subject, and visualise what actually is here the author's intention. How did diversity arise? How could here be a development of the distinction between the seer and the seen from that theoretic nebular condition of universal darkness and cosmic waters? That condition is not of the Absolute, but what sometimes is described in the Purānas, and in the Epics, as the precondition of the manifestation of the external universe. It is difficult to imagine this condition, because we cannot understand what could be the precondition of the manifestation of externality, which is what we call creation. Creation is nothing but the projection of externality in Indivisible Being. The creation of the universe, therefore, is not actually the manufacture of a new substance. This is the great point which will be explained in greater detail, further, as we proceed.

In creation, a new thing is not created, because nothing can come from nothing. If a new thing is to be created, it must have been produced out of nothing. How can 'nothing' produce 'something'? (Sankhya Sutra 78) "A thing is not made out of nothing" This is illogical. The effect must have existed in some causal state. This causal state is the substance of the universe. Now, what is actually the distinctive mark of the universe that is created, as different from the original causal condition? In what way does the effect get differentiated from the cause? If everything that is in the effect is in the cause, what is the distinctive feature, what is the distinguishing mark, which separates the effect from the cause? If the effect is entirely different from the cause, we cannot posit a cause at all, because the cause is non-existent. If the cause is non-existent, the effect also would be non-existent. So, the cause must have contained the effect in a primordial state; and, therefore, nothing can be visualised in the effect which could not have been in the cause. In a sense, therefore, what is in the effect is what is in the cause. The effect is the cause. There is no final non-distinction between the effect and the cause, inasmuch as in substance they are the same. But yet, we make a distinction between the two.

This peculiarity, Višeshata, which characterises the distinction between the cause and the effect, is the principle of what we call space-time in modern philosophical language. But, otherwise, it is the principle of externality. The principle of externality is not a substance. It is a peculiar state of consciousness. That is the distinguishable principle. The effect gets isolated from the cause by a peculiar adjustment of consciousness within the cause, not necessarily involved in the change or modification of the cause, but only a state of mind or consciousness. Now, when the effect gets psychologically isolated from the cause, there is the seed sown for the further diversity of creation. The two become four, four becomes eight, eight becomes sixteen, and multiplicity, thus, proceeds from the original Single Atom of the cosmos. And, when this diversity, which is creation, is conceived as possible and capable of being hiddenly present in the cause, we have to assume, also, a peculiar potency in the cause, which becomes the reason behind the manifestation of diversity. This is called the Šakti in certain philosophies, the force, energy, that is present in consciousness, a peculiar indistinguishable, indescribable, and eluding something, without the assumption of which creation cannot be assumed. And, sometimes, people call it Māya, merely because they cannot understand what it is. It is not a substance that exists. It is rather an inability to grasp the meaning of it; that is all.

Now, this peculiarity, whatever we may call it, whatever designation may be applied to it, is the cause of the distinction of the effect from the cause, and that becomes the first breeding ground for the further multifarious division we see in the form of this vast creation. The moment this creation begins, the moment there is the potency released for the external expression of what was hiddenly present in the cause; there is a catastrophic change taking place. And, this is the urge for creation, the urge for diversity, multiplicity, colour, sound, activity, etc. This characteristic of self-division is called Mrtyu (death priciple), that which destroys the indivisible, that which isolates the one from the other, that which disfigures the original condition of things, the destroyer of the original state of affairs. That is symbolically called death here, and further, it is described as the hunger of things to grab other objects.

Now, what is this hunger mentioned here – *ašanāyayā hi mrtyuh*? It is the urge that is simultaneously present in the process of creation for an involution of things. When there is a separation of one thing from another in creation, the seer becomes distinguished from the seen, the subject is separated from the object, they struggle to become one; because that which is separated has hiddenly present in itself the capacity to unite also, as the two are nothing but the substance of the one. So, the indivisibility of the one presses itself forward even in the divisibility of the two. So, there is restlessness everywhere. Our sorrows, our difficulties or problems, our griefs and every kind of unwanted things here, are a tussle between two elements in our soul – the urge for diversity and the urge for unity, fighting with one another. This struggle is Samsāra, right from the original Creator, Brahma, down to a blade of grass. This Ašanāyayā, the hunger of the spirit, is the activity of the cosmos, where, on one side, it struggles to become more and more wide in its physical quantitative expanse, and on the other side, it struggles to become one with the Universal Spirit. So, we have two elements present in us always – the tendency to unity and the tendency to diversity. We ask for expansion in quantity, and at the same time, we ask for a heightening of our value in quality. However, the Upanishad here mentions, in a very difficult word, that the origin of creation is indescribable, and it is indescribable merely because it preceded a state which requires the presence of the effect in the cause, and which was also preceded by a state which has within it, invisibly present, the capacity to multiply and also the capacity to unite.

The mind of the cosmos, which is called the Cosmic Mind, in usual parlance, is regarded here as an evolute, and not the original Being. The Absolute is Transcendent Being, and not a mind, thinking. It is not even a casual state. Even the casual state is supported to be posterior to the Absolute. We never associate the Absolute with the world. The Brahman of the Upanishad, or the Absolute of philosophy, is the assertion of being which is unrelated to creation. And, when we have to associate God with creation, we have a new word altogether for it. Īshvara is the word we use in the language of the Vedānta. Such words do not occur in the Upanishads. They are all to be found in the later Vedānta, but they are assumed here.

In the Sāmkhya and the Vedānta cosmological descriptions, we have certain grades mentioned of the coming out of the effect from the cause. Before we go further into the difficulties envisaged in these passages of the Upanishads, it is better to understand the evolutionary principles as initiated in the Sāmkhya and the Vedānta. The Sāmkhya tells us that there was an original condition where everything was potent, though not patent. Everything was hidden, though not expressed. Everything was in a universal causal state. That is regarded as the non-existent, dark, undeveloped, indivisible state of things. That is called Prakriti in the Sāmkhya language. Those of us who have studied the Sāmkhya philosophy will know what is Prakriti, and how evolutes proceed, come out, from this Prakriti. Prakriti is only a Sanskrit term for the matrix of all things, the original state where everything is in a mass, where one thing cannot be distinguished from the other, what the astronomers would call the nebular dust, in some way. But this is something more than that. It is a cosmic death, one may call it. Everything is contained there, and everything is hidden; everything is undeveloped and indistinguishable, incapable of being perceived, because even the sense-organs are not developed there.

Then, there is a tendency to think. The cosmic thought develops itself. That is what is indicated here by the words, 'tan mano' 'kurata'. From this undeveloped Being which was equivalent to universal darkness, mind arose. That mind is the Cosmic Mind. In the Sāmkhya, we call it Mahat; and in the Vedānta, we call it Hiranyagarbha. This cosmic undeveloped state is sometimes called Īshvara. Now, Īshvara is not undeveloped in the sense of a primitive state where intelligence is absent, but it is an exceedingly intelligent condition where distinctions are not present. We call it symbolically dark, because the light of the senses will not operate there. It is a light that is transcendent; and in the passages occurring in such verses as the Manusmriti, we are told that it was shining as brightly as thousands of suns, Sahasramśusamaprabhm. How can we call it darkness? But, it was darkness to the eyes which were not developed, just as the blaze of the sun be darkness to the eyes, when it is very intense.

So, the mind that is supposed to be the evolute, immediately proceeding from the undeveloped condition, is the Hiranyagarbha principle of the Vedānta, coming from the Īshvara principle, or Mahat coming from Prakriti. Then, there is the Ahamkāra proceeding from Mahat, the Self-sense of the cosmos. This is how the Sāmkhya would describe the development of the original, Cosmic 'I'-sense from the Cosmic Intelligence, which, again, is an evolute of the Cosmic Prakriti. Then, there is the distinction between the subject and the object; on one side, there is the physical universe, and on the other side, there are the individuals. The physical universe is constituted of the Tanmātras

– Šabda, Sparša, Rupa, Rasa, Gandha, which become concretised by a process called quintuplication into the five elements – ether, air, fire, water and earth. And, subjectively, they become the individuals with the five Koshas – Annamaya, Prānamaya, Manomaya, Vijñānamaya and ānandamaya. These Koshas are the vestures of the individual soul – the physical, the vital, the mental, the intellectual and the causal bodies. These are called the five Koshas. And within these Koshas we have the Prānas, the senses of perception and action, and the mind, the ego, the subconscious, the unconscious, and the intellect; and ultimately, a very unintelligible substance within us which we experience in deep sleep – that is the causal state. So, this is how the Sāmkhya would describe the process of creation, which is followed literally, to some extent, in the Vedānta also, with only a distinction in definition. Instead of the terms; Prakriti, Mahat, Ahamkāra, we have the terms; Ìshvara, Hiranyagarbha, Virāt.

So, this cosmological process, the development of the effect from the cause, gradually, from the Universal Being, down to the lowest of diverse elements – this it is that is described here in this Brāhmana, which says that originally nothing was, from where the element of distinction between the subject and the object, characterised by a double activity of grasping and separation, was evolved, and then arose the Cosmic Mind, Hrianyagarbha.

Here is a passage of great significance from the point of view of philosophical technique employed in the understanding of the relation between the individual and the Universal. This which is a symbolic statement in the Upanishad, very hard, indeed, to understand, conveys a wealth of meaning. What exactly is the connection between the diverse individuals and the Universal Absolute? This has been a great point of discussion throughout the history of philosophy, and it is not easy to come to a conclusion. Often, it is thought that the Universe is a collection of all the individuals or particulars. Many a time, we are told by philosophers that the Absolute is the whole, and the individuals are the parts thereof;

so that to get the absolute, one has only to collect all the individuals and group them together, which means to say, anything that we find in the individual will be found in the Absolute. There will be nothing more in the Absolute than what we see in the individual. This conclusion also will follow, if this assumption is correct; and it is a very uncomfortable conclusion, because we are not seeking in the Absolute merely what is in us. A million people put together cannot be regarded as qualitatively superior to what a single individual is. It is also held that the Absolute is transcendent in the sense that it has no connection at all with the visible universe. Often, it is also held that the Absolute is so much absorbed in the universe that we cannot find it outside the universe. So, we have theories and theories, and doctrines and doctrines.

This Upanishad, in this one single sentence, tells us what the fact is. The original condition, causing the manifestation of diversity, is the death of universality. This is what is called Mrityu. The death of something becomes the birth of something else. For the birth of the individual, the universal has to die. Very strange, indeed! We cannot understand what this means. The death of the universe means the complete abolition of the consciousness of the universal; and for all practical purposes, death and absence of consciousness are the same. The condition that is requisite, absolutely necessary, for the manifestation of the universe in the form of diversity, is an abolition of the consciousness of the Absolute, because there is no question of the manifestation of diversity in the Absolute. Manifestation requires space, time and cause, and many other things that follow. If the Absolute is spaceless and timeless, durationless, infinitude, eternity, the question of creation, manifestation, etc. does not arise there. Then, how comes this universe? From where has this universe arisen, or the diversity come? It can be explained, says this Upanishad, by a strange phenomenon that should be assumed to have taken place, it as all creation is to be taken as a fact.

The consciousness of the existence of the universe is different from the consciousness of the Absolute. That the two are not identical is a point that is made out here. Once the existence of the universe is accepted in consciousness, everything else that follows from it can also be accepted. If two and two make four, four and four make eight, and so on, arithmetically, we can draw conclusions. But two and two must, first of all, make four. We must accept that. If that is not true, then any multiplication, therefrom, also is not true. There is a distinction between Absolute-Consciousness and universe-consciousness. That distinction is the cause behind this line drawn here between Pure Being that is Absolute, and the condition precedent to creation. It is difficult for the human mind to understand what the Absolute is. Whatever be our stretch of imagination, we cannot conceive it, because every conception is quantitative and qualitative. The Absolute is neither a quantity nor a quality, and therefore no thought of it is possible. Even the subtlest thought that can be applied to the Absolute is, after all, a magnified form of the quantity-quality relation in terms of which alone is the mind able to think. There is no such thing as 'thinking' the Absolute. Such a thing is not possible, because the thought which thinks the Absolute cannot exist independent of the Absolute; for, what we call the Absolute is that which includes everything, including even the mind. So, the mind that thinks the Absolute is a part of the Absolute itself, and therefore the mind cannot think the Absolute. This is a very reasonable conclusion. Inasmuch as the thinker is involved in what is thought, there is no such thing as thinking at all in terms of 'That.' Either the Absolute is outside the mind, in which case it ceases to be the Absolute, or it is not an object of thought. It is not even a concept for philosophical disquisitions. But that being the nature of the Absolute, we cannot attribute to it any quality that is visible in the universe of creation. What about diversions, three dimensions, for instance? The three-dimensional universe, which is of space and of time which is duration, cannot be correlated with the Absolute, if this is its character, this is its nature, and this is the essence of its Being.

In order that the universe may be manifest, some phenomenon should take place; and that phenomenon is described here as Mrityu.

And Mrityu, here, does not mean the ordinary phenomenon of death or destruction of a body. It is a metaphysical concept that is introduced here. It is a tentative withdrawal of the consciousness of the Absolute, and a manifestation of a new universal which embodies within itself, in a seed form, everything that we call the gross universe. The Will of God is supposed to be the originator of the universe, as we hear of, as proclaimed in the scriptures of the religions. The God of the universe, who is the Creator, manifested through His Will all this creation. Now, the attribution of 'Will' to God is indeed a difficult task, because, as far as we know, Will is a psychological function, and it can be defined in certain specific manners. But the definition of the 'Will' that we have in psychology is something which cannot be attributed to a God who is Universal. However, we have to assume a different kind of 'Will,' and the Will which is responsible for the projection of the universe in a seed form, originally, can be described as a kind of potency or potentiality or latency of being, as the seed may be said to be the latency of the tree. The vast banyan tree which is so big, grows towering to the skies, is hiddenly present in a very tiny seed, as we know. We may say that the seed is the potential condition of the tree, though if we cut the seed, we cannot see there anything of the tree. Visibly, there is nothing; but we have to infer the presence of all the diversity of the banyan tree in this little seed which is so tiny. Likewise, a condition is assumed which is the potential seed of all the diversities to be manifest.

Now, many thinkers of the topmost calibre, in the field of philosophy, have held that the cause of creation is not a desire on the part of God, as many would ordinarily think, because it is impossible to imagine that God can have a desire. Achārya Šankara, and such other thinkers, tell us that the cause of the universe is not the desire of God, just as the moon shining in the sky is not the cause of the thief breaking into somebody's house – a very beautiful analogy. If, with the help of the moonlight, some burglar enters somebody's house, the action of burglary cannot be imputed to the moon because it is responsible, in some way, in shedding light to the thief. Likewise is the presence of the Will of God in the process of the manifestation of the universe. The activity of creation, or the substance, the material of creation is, in some way, distinguished from the efficient cause of creation. The efficient cause of creation is the potency of God's Will, which does not desire the world to be created, but becomes necessary for the manifestation of the universe in a particular fashion. The fashion, the pattern, or the shape which the universe takes in a particular cycle of time, is supposed to be the grossened form of the subtle psychological or psychic potency, present in the individuals who lay unliberated at the end of the previous cycle, or the Kalpa, as we call it. The individuals who are not liberated at the end of the world lie potent, latent, seed-like, in the bosom of the cosmos, and they are said to lie for as long a time as the universe lasted earlier. Such is the night of Brahma as the scriptures tell us, as was the day of Brahma, earlier. (The reader will get to know clear picture once he has read Manvantara Theory of Evolution).

The night of the cosmos is compared to the cosmic waters in some mythologies, as we have the waters mentioned, also, in this Upanishad. The cosmic waters, mentioned in creative or cosmological theories, are nothing but the original condition of things, subsequent to the dissolution of the cosmos, and prior to the creation later on, during which period the unliberated individuals lie like seeds about to sprout. A particular set of individuals-they may be millions, hundreds of millions, thousands of millions, etc. – are grouped together in a particular category; and this grouped category of individuals, in their generality of psychic structure, becomes responsible for the material shape which the universe has to take after the fructification of those potencies. Just as the seed does not sprout into a tree at all times – it requires conditions, such as proper atmosphere, good climate, rain and suitable soil, etc. – the individuals who are lying in a seed form do not sprout into activity until maturity takes place. This maturity is supposed to take place somewhat like the waking of the individual from sleep. How long do you sleep in the night?

As long as it is necessary for the psychic potencies to wake up into activity. The awakening of the psychic potencies within, into activity is called waking from sleep, which happens into the daylight of consciousness. Something like that is supposed to take place, cosmically, during the time of creation. The individuals, collectively, feel the fructification of the psychic contents and they germinate into action. And, the world that is manifest, the universe that is projected is of a character which is necessary for the fulfilment of the desires left unfulfilled by the individual during the time of the dissolution of the universe earlier.

So, such is the very interesting doctrine propounded by thinkers like Achãrya Šankara. We find it in the Brahma-Sútras, especially, mentioned in a very concise form. Perhaps, the doctrine is based upon the Upinishads, which are more concise and less clear in their exposition. Here we have such a type of doctrine of creation, which makes out that the consciousness of the world is the reverse of the Consciousness of the Absolute, which is very strange for us to hear and even to understand. It is not a part of the Absolute that we are seeing when we are looking out into the world. We are seeing something topsy-tyrvy, a reversed form, as we see ourselves reversed in water as a reflection. When we stand on the bank of the Ganga and see ourselves reflected, we will find that the head which is topmost will be the lower most there. The feet which are the lower most will be the uppermost in the reflection. So, there is a complete reversal of the position of the body when it is reflected. Some such thing is said to have taken place at the time of creation, so that, when we see the head of ours reflected in water, it appears to be our head, but it is not really our head. The head that we see, reflected in water, looks like our head, and is exactly like our head. We may mistake it for our head but it is not our head, really. Likewise, we may mistake these things of the world for the Absolute, but they are not, in the same way as we may think the reflected is ours, but it is not.

And, also, another analogy is given in a passage of the Katha Upanishad as to what happens in creation. There is a reversal of the whole position, as our face is reversed in a mirror where it is reflected – the right becomes left, and the left becomes right – even so, the subject becomes the object, and the object becomes the subject, when the creation takes place, which is the essence of the whole matter. Very interesting and very comforting indeed! We can imagine where we are seated, and what has happened to us. God has played a very beautiful joke with us, made us great fools, turned us upside down by positing the subject in the context of the object, and the object in the context of the subject. Really, we are the objects; the universe is the subject. This is the truth. But, we think that we are the subjects, and the universe is the object, and gaze at it, look at it, try to exploit it for our own individual purposes, under the misapprehension that we are the subjects. We are subjects in the same sense as the reflected head is our head.

So, this reversal of the position of the Absolute is called Mrityu, or destruction, or death, here. Well, it is destruction indeed, when we mistake one thing for another thing by completely forgetting the original, and we are destroyed, in fact, when we are in a different paradise altogether, where we are under an illusion. And, consciousness gets reflected where ever there is this reversed position, cognised or felt, where consciousness attends. So there is a reflected consciousness, also. The entire personality of ours maybe said to be a reflected structure. Even the intellect is a reflection of the consciousness of God. It is not qualitatively equivalent to God-consciousness. It does not mean that a tiny part of God is in our brains. Not so; it is reflected, which means distorted. The sun, reflected in water, may look like the sun, but it does not have the quality of the sun. It will not burn you. You cannot warm yourself by the reflected sun in the water.

There is diversity in the form of this creation, made possible by a reversal of the position of the ultimate Reality and that reversed position assumes a consciousness of its own, originality. That is what is known as the Universal Mind. It is attended with Self-Consciousness – **ātmanvī syām iti**. 'I-Am', the cosmic 'I-Am', is something less than the Absolute.

It is a condition that has to be accepted subsequent to the reversal, which, again, has to be assumed prior.

The Cosmic Mind, Hiranyagarbha, as we call it in the Vedānta, is the Cosmic 'I-Am.' It is Self-Consciousness, Pure Universality. And, here is the seed of all diversity. In a sense, we may say that we are parts of this Cosmic Mind, but not, indeed correctly. As I pointed out, we cannot regard ourselves as parts of the Absolute. Nothing that we see with our eyes can be regarded as a real representation of the Absolute. Thus, we have to understand that we are not parts, even of the Hiranyagarbha. We are much less than that. We are far down below the condition of Hiranyagarbha and Virāt, for reasons we shall see shortly. For the time being, it is enough if we understand the actual meaning of this passage. There was a destruction, a Mrityu, a complete abolition of Reality, which is what the Sāmkhya calls Prakriti, the Vedāntins called Maya, Mula-Prakriti, etc., the Potential Being, the Matrix of the universe. That becomes the seed for the manifestation of the Cosmic Mind, known as Mahat and Cosmic Ahamkāra. The Vedānta calls them Hiranyagarbha and Virāt.

So'rcann acarat, tasyārcata. āpo'jāyanta, arcate vai me kam abhúd iti; tad evārkasya arkatvam; kam ha vā asmai bhavati, ya evametad arkasya arkatvam veda. One who makes this phenomenon, assumes power over this phenomenon, becomes that, is the advice with which the passage concludes. The cosmic condition is thus to be described. The Mind which was created, cosmically, in this manner, by a reversal of the content of the Absolute, this cosmic condition is the seed of the universe. This seed of the universe we call Īshvara; we call Hiranyagarbha; we call Virāt, in the various degreesof the densities manifested. It assumed a joy. It became the energy of the universe. It became Vaisvānara. Here the word 'Arka' is sometimes taken to mean Cosmic Fire or may say, Universal Energy, whish is aslo the same as the great Joy of the Universal. 'Kam' means joy, happiness. There is a happiness which is untarnished and undiminished in this condition on account of the retention of universality, though it is the seed-form of all diversity. The conditions of Hiranyagarbha and Virāt are potential diversities, no doubt, but not manifest diversities. What we call diversity, responsible for the sorrow of the individual has not taken place yet. There is no sorrow in Virāt, Hiranyagarbha and Īshvara, though there is a potency of diversity. The reason is that there is the Universal Consciousness maintained yet, inspite of the potentiality for diversity. There is an organic connectedness of things in Virāt and above, and this consciousness is maintained. Therefore an account of the absence of the loss of universality there, the Joy of the Universal also is present. Whoever knows this becomes that; and knowing this, he is also equal to have the power of it. Knowledge and Power are identical. So the knowledge of it is necessary, by 'being it' in meditation. And then there is power that is unlimited, power that is born of unlimited knowledge on account of unlimited 'Being.'

1. āpo vā arkah. tad yad apām śara āsít, tat samahanyta, sā prithivy abhavat, tasyām aśrāmyat. tasya śrāntasya taptasya tejo raso niravartatāgnih.

Here, again, we have some description of the condensation of the dense form of things, gradually taking place in the process of creation – the subtle becomes gross. The cosmic waters hardened, as it were, became solid, gradually, and the Earth element was formed. By the Earth element, what is meant here is not merely this little globe of the earth on which we are living, but the entire Earth Principle of the whole astronomical universe, through which your eyes cannot pass. The whole element of Earth can be regarded as the solidified form of this cosmic condition, the subtle nature of things which is called here, Waters. It solidified itself. From Fire comes Water; from Water comes Earth. This is the chronological order of creation of the gross forms, ordinarily speaking. *Sā prithivy abhavat*: That became the Earth, the grossened form of things.

Here is the end of cosmic creation. There is a famous passage in a text of the Vedānta, known as Panchadaśi, written by Sage Vidyāranya, who describes this in one Sloka: *Ikṣānadi-praveśāntā srshtirīśena kalpitā. Jāgradādi-vimokshantah samsāra jivíkalpitah.* This passage of the Upanishad and such other passages are given their meaning in this verse of the Panchadaśi. From the Cosmic Will down to Divine Immanence, it is Īshvara's creation. From walking till liberation, it is the individual's creation. Īshvara's creation or God's creation ends with the manifestation of the universal physical form, and God is not responsible for what the individual is experiencing. The loves and sorrows, the joys and pains, the births and deaths of the individual are not created by God. They are created by some other factor which is not to be attributed to God. The condensation of the cosmos, right from the causal condition down to the physical, through the subtle, may be said to be the manifesting activity of God. He becomes the 'All' and becomes also the consciousness of the 'All.' But the reversal of attitude, the considering of the object as the subject and the subject as the object, and the desire to grab objects for the purpose of personal satisfaction, and the capacity to fulfil certain desires and the incapacity to fulfil certain others, the getting fatigued in personality on account of the inability to fulfil all desires, the falling into sleep every day on account of the latent condition of desires unfulfilled, etc. – these are all the phenomena of individuality, not of Cosmic Being. Even the 'process' of Moksha, or liberation, is not God's creation, because God has no Moksha. He is always in the state of Moksha only. The process of bondage and liberation, the cycle of births and deaths and joys and sorrows and activity, everything of this nature is an outcome of certain subsidiary character assumed by the individual, isolated from the Universal, so that we may say that there is no sorrow down to the point of the Virāt manifestation. Sorrow starts after that, when there is a split into the diverse individuals who regard themselves as self-contained, self-sufficient, self-exhaustive individual. Each one of us regards himself as complete. That there is nothing lacking in us, is a misconception. We lack everything, but we think we are complete in ourselves, so that we have a soul of 'our own,' an entire soul, which is entirely ours, independent, unconnected with others! This is called the ego-principle which affirms a total isolation of itself from others. This has happened subsequently, and anything that follows out of it is the responsibility of the Jīva, the individual, not of Īshvara.

Here we have a description of creation down to the point of Virāt. *Tejo raso niravartatāgnih:* A luminous essence, which we may call the Cosmic Fire, emanated from this condition, which is the outrush of the Creative Process. That luminous Cosmic Essential Being, the Fire Universal, is what we call Vaiśvānara or Virāt. Then what happens? We are slowly to come down to our sorrowful state, not yet begun, but going to begin.

The intermediary conditions are now described, which are prior to the manifestation of our grossened individualities. There are certain intermediary stages – the division of the Virāt into the Tripartite Being, known in technical language as Adhyātma (subject), Abdhibhúta (object), and Adhidaiva (transcendent). There is no such thing as Adhyātma, Abdhibhúta, Adhidaiva in the Virāt. All the three aspects are one there, but these three have to be separated and conceived independently for the purpose of subsequent creation. That point is slowly being arrived at, in these passages.

1. sa tredhātmānam vyakuruta, ādityam trtíyam, vāyum trtíyam, sa esa prānas tredhā vihitah. tasya prācí dik širah, asau cāsau cairmau; athā asya pratící dik puccham, asau cāsau ca sakthyau; daksinā codící ca pāršve, dyauh prstham, antariksam udaram, iyam urah, sa eso'psu pratisthitah, yatra kva caiti tad eva pratitisthaty evam vidvān.

Threefold is the manifestation subsequent to this original condition. *ādityam trtíyam, vāyun trtíyam sa esa prānas tredhā vihitah:* Here Prāna means the Cosmic Prāna, Hiranyagarbha, or we may say, Virāt. He assumed a threefold form – the transcendent (Adhidaiva), the objective (Abdhibhúta) and the subjective (Adhyātma). Prior to this, there was no such distinction as the transcendent, the objective and the subjective. Now we have the God who is above, the world which is outside, and ourselves here. This tripartite distinction has now taken place. So, when we pray to God, we look up as if He is 'above.' He was not above previously. Now He has become above because we have lost Him. He has run up to the skies, as it were. And the world is 'outside' us, and we are looking at it, and we are 'here' as imagined subjects. We are subjects falsely arrogated to ourselves. This is, perhaps, the fall described in the Biblical context, the Satan falling (Isaiah 14: 12-20), assuming individuality, independent of God. The assumption of individuality immediately calls for a transcendent Creator and an external universe. The moment you become conscious of yourself as an isolated being, you begin to see an outside world, and then you conceive, not perceive a transcendent God. Here, God becomes merely a conception; He is not an object of perception. Originally, He was a content of direct perception, experience, realization; He was 'Being,' 'Existence,' 'Vitality,' the 'Soul' itself. Now He has escaped our grasp, and over and above us become transcendent, and remained only as a theoretical Creator for our prayers and worship. What we physically see is only the world of gross objects, towards which we run every moment of time, assuming that we are the sole monarchs of the world, that we are the rulers of things; an assumption, false indeed, for reasons quite obvious.

This Cosmic Prāna, Hiranyagarbha, or Virāt, assumed a threefold aspect – Adhibhautika, Adhyātmaika and Adhidaivika, viz, the physical, the subjective and the transcendent. The objective or the physical, the subjective or the psychic, and the Transcendent which is the invisible divine content, are later formulations.

Here again the Upanishad brings us back, by a Simhāvalokana, as it were a retrospective look, to the unity of things, in spite of the tripartite diversification that has taken place. In spite of this threefold manifestation, which is apparently a segmentation of creation into three different corners, as if unconnected with one another, there is yet a unity among them. That point is brought out here, in this analogy, which describes the unity present in the midst of this tripartite diversity, by the comparison of this triad with that of the horse in the Ašvamedha Sacrifice, and also in terms of a particular shape of the sacrificial ground takes in the Ašvamedha Sacrifice, viz, the shape of a bird. The sacrificial ground is drawn in a particular shape. The shape is of a bird. So, the bird is described here, or we may say, the horse itself is described. Both comparisons are apt. The eastern direction of this sacrificial ground in this drawing which is the shape of a bird, or of this Ašvamedha Sacrifice horse; of this, the eastern direction is the head. And the various limbs are described further, as before. If arms are the intermediary quarters, northeast and southeast. The western quarter is its tail. Again, the hip bones in the body of the horse are the other intermediary quarters, viz, northwest and southwest. The southern diection and the northern directions are the sides of the body. The sky is the back; the atmosphere is the belly; this earth is the chest. And this is the description of the cosmic condition. This Virāt description is to be found in the sacrificial diagrams of the Ašvamedha Sacrifice, as also in temple constructions.

The temples, especially in Southern India are constructed in the shape of the Virāt. The Holy of Holies inside is the head of the Virāt, which is represented by a luminous glow of a sacred light in a dark room, comparable to the ānandamaya Kosha (causal sheath) which is dark, but illuminated by the ātman within, and encompassed by seven Prakaras, or corridors. Sometimes these are five, comparable to the five Koshas or vestures of the body – Annamaya, Prāṇamaya, Manomaya, Vijñānamaya, ānandamaya – the physical vital, mental, intellectual and causal sheaths. And there is the Balipitha, the sacrificial altar, at the entrance, which is represented by a huge post. Before you enter the body of the Virāt, you have to offer yourself first; otherwise, no entry is possible. You have to pay a fee to the Virāt before you gain access into it, and the fee is your own self. You have to cease to be, first, as you are now, in order that you may become what you want to become. This is a symbol of temple construction, and also of the patterns drawn in the

The Creation of the Universe – The Brihadaranyaka Upanishad

Aśvamedha Sacrifice. That pattern is described here in the correlation with the parts of the universe. Such is the geometrical description of the creation of the universe, with its deep philosophical significance and spiritual connotation. One who knows this becomes strong and obtains a resting place wherever he be.

1. so 'kāmayata dvitíyo ma ātmā jāyeteti, sa manasā vācam mithunam samabhavad aśanāyā mrtyuh, tad yad reta āsít, sa samvatsaro 'bhavat; na ha purā tatah samvatsara āsa. tam etāvantam kālam abhibhah. yāvān samvatsarah tam etāvatah, kālasya parastād asrjata; tam jātam abhivyādadāt, sa bhān akarot saiva vāg abhavat.

It willed, or He willed: "May I have a second Self." This is the origin of creation. The world, this creation, this universe is the second Self, as it were of the Supreme Being. This 'other' Self, which is this vast creation, is animated by the Supreme Being Himself. It is 'other' in the sense that is had not all the characters of the Absolute. Yet, it is the Self. Though it is the 'other,' it is also the Self; it is called the 'Other Self,' inasmuch as the Selfhood of the Absolute is transparently present in this creation. The Universal atman is immanent in the whole universe, in all aspects of creation; and yet the universe is an 'otherness,' as it were, of God, an object of God. It is as if the Universal 'I' is envisaging a universal object, including all that is visible or sensible – space, time and causal relation. A single Subject encountering a single Object is a state which is described in this passage, a Cosmic Consciousness becoming aware of a Cosmic Object in a peculiar manner, not in the way in which the ordinary individual is aware of an object outside. The way in which God is conscious of the universe, is different from the way in which an ordinary Jíva, or individual, is conscious of an object. This makes all the difference between Universal Consciousness and particularised consciousness.

The object, in an ordinary perception, is segregated from the subject by the differentiating medium of space and of time, so that there is no vital connection between the object that is perceived and the subject that perceives. But there is a living connectedness between the Cosmic Object and the Cosmic Subject. This connection is sometimes described as one of Body and Soul. We know that there is a connection between the soul and the body. This relation between the soul and the body is different from the relation between an individual subjects encountering an outside object. The soul and the body cannot be separated from each other. They are organically one. This relation is called Šariri – Šariri-Bhāva, the relation between consciousness and its embodiment. Thus, we can say that Cosmic Awareness of the universe, in the case of God-Consciousness, is one of inseparable relation, like the relation of the soul and the body. When we are aware of our bodies, we are not only becoming aware of an object situated in space and time. We can say that this bode is an object because it is sensed, it can be seen, and it has all the characters of any object in the world; but, at the same time, it is an object which clings to us vitally and organically, not like an object such as the mountain which is far off in space and, perhaps, in time.

There are three kinds of 'self' distinguished in the philosophy of the Vedānta. These three 'selfs' are the three aspects of the conception of the One Self. They are called the Mukhya-ātman, the Mithya-ātman and the Gauna-ātman, in Sanskrit. The Mukhya-ātman is the primary Self, which is uniform and unique in every individual, equally. It does not differ from one person to another person, from one thing to another thing, like space contained in various vessels. It is the same space that is in all vessels, irrespective of the number and size of the vessels, etc. This ubiquitous Consciousness, which is equally present in all beings, irrespective of the distinctions in space, time and cause, is the Absolute Self. That is called the Mukhya-ātman. There is also the Gauna-ātman, or the secondary self which is distinguishable from the primary Self. It is not merely that one has within oneself, immanently present, the eternal primary Self, but there is also another kind of self with which one's individuality is connected. Anything that one loves is also a self. As a matter of fact, all love is a movement of self in a particular direction.

When the self moves, we call it 'love'; and when it does not move, we call it 'being.' But, it is the same 'self' that acts, whether it moves or whether it does not move. The movement of the 'self' towards an object for any particular given purpose becomes the cause of affection for that object, and the 'Self' which is primary, is recognised in the object which is secondary. So, in the love of the object we are loving our own self really, it is not just something else. The object is invested with the character of the 'Self,' and then there is an immense affection felt for the object. Every form of love is the love of 'Self.' There is nothing else in any form of affection. The object which is thus invested with character of one's own Self, becoming the centre of affection, is what is called the secondary self. It is also a self, but it is not the Absolute Self. So, it is called secondary. The third form of self is this body which is temporarily assumed as the 'self' for the purpose of working out certain Karmas done in previous births. The nature of this body is characterised by the structure of the desires expressed in previous lives, and the Karmas performed in previous lives. Karma, or an action, is a desire that is externalised in respect of an ulterior motive. Every action is desire-propelled. A desire-propelled movement in the direction of an object is an action, and that action produces a reaction, because every action is an interference in the universality of the cosmos. The equilibrium of the universe is disturbed by every action of an individual. This disturbance that is caused by the action of an individual is set right by the balance that is ever maintained by the universe. And this balance is maintained by a reaction that is so set up. The reaction comes back as a boomerang upon the very individual that is the source of that disturbance. This is called the Karma-Phala, or the fruit of action. That Karma-Phala becomes the seed for the manifestation of a future body. So, this body which we are assuming today, and in which we are embodied, is the result of our past Karmas. It is of such nature, such a character, such duration of life, etc., as were our previous desires and actions. This body also is an atman for us. We love it immensely. So it is 'self,' but it is a 'false' self. It is not the real Self.

So it is called the Mithya-ātman. Thus the threefold distinction of the atman is made in this manner-the Mukhya-ātman, the Gauna-ātman, the Mithya-ātman-the primary Self, the secondary self, and the false self. Here, the Universal Being Willed "Let me have a secondary Self." This is perhaps, the meaning of this passage of the Upanishad.

You have heard this great passage of the Bible: In the beginning was the Word, and the Word was with God, and the Word was God (John 1:1). Something like this is what the Upanishad tells us here. The Eternal Wisdom was manifest, with the eternal Word, and with this Word the whole cosmos was created. The Word which is with God, and which is God, is not merely a letter, or a sound that we make through our lips. It is energy; it is a force; it is a vibration, which materialises itself, concretises itself into object-forms. The Word is the Veda, or Eternal Wisdom which is with God, and it is inseparable from God, and so, it is God Himself. The cosmic Mind projected itself in the form of this Eternal Word, and manifested this universe. In the Manusmriti, and such other ancient texts, we are also told in a symbolic manner that Prajāpati, the Creator, conceived the whole cosmos in the pattern of 'Om,' or the Pranava. The Pranava, or Omkāra, is supposed to be the seed of the whole universe. That is the essence of the Word that is Divine. It is also the Veda contained in a seed form. The whole of the Veda is inside 'Om.' "Eka eva purā vedah pranavah sarva-vāngmayah," says Bhagavan Sri Krishna, as recorded in the Srimad-Bhagavata, when he spoke to Uddhava. There was only one Veda in the beginning. It was 'Om.' We did not have four Vedas like Rik, Yajur, Sāman and Atharva. They were classifications made later on by Sage Veda-Vyāsa. 'Om' is supposed to be a vibration, which is integral in its nature, and that is the Word spoken of. This Word which id Om, is the cause of the whole cosmos. The Mind of the Universe, the Cosmic Mind, Prajāpati, got united with this Word, which means to say, Consciousness vibrated through this Word for the purpose of the manifestation of the universe. And, in the Manusmriti, we are told that Pranava splits itself into the Vyāhritis-Bhúh, Bhavah, and Svah. These are mystical syllables which are supposed to contain the inner content of the Pranava. And we are also further told that the three Vyāhritis split themselves into the three Padas, or the quarters of the Gāyatri Mantra which is supposed to be expounded in a greater detail in the three sections of the Purusha-Súkta. These three parts of the Purusha-Súkta become the three Vedas-Rik, Yajur and Sāman, and in all three multiplications. So, the origin of this creation is supposed to be a communion of the Cosmic Mind with Cosmic Vibration, which is referred to as the Word, the Veda-Vac, which means speech, the Original Word.

Sa manasā vācam mithunam samabhavad ašanāyā mrtyuh: Here the word ašanāyā mrtyuh is repeated once again in order to bring out the sense that creation is an 'othering' of God, an alienation, a sacrifice, which is sometimes called the 'Cosmic Sacrifice.' The Absolute becomes something other than itself, in order that it may appear as this universe. How does it become other than it is? By the projection of the time factor. There is no time in God; it is all Eternity. The moment there is the projection of process, it becomes creation-Samvatsara, the time-cycle. Samvatsara is the principle of the year, which is time. The moment there is consciousness of time, we are in a world of experience. And in the Absolute, which is durationless Eternity, there is no such process as time; there is no past, present and future. What we call Eternity was the Essence of God Himself, and in the grasp of the Universal Consciousness of God, past, present and future come together in a single comprehension. But, in the individual's case, this is split into three sections-the past, the present, and the future, which cannot be connected easily. We cannot know the past, we cannot know the future, we are in a very fine split-fact of what is called the present. Every second, the present passes and becomes the past. The past, the present, and the future are not three distinct parts of time, cut off one from the other. They are continuity like the flow of a river. But, due to a peculiar effect that the time has upon our minds as individuals, we are unable to conceive of the past and the future, and we are stuck up in the middle, in the present merely. However, the point made out here is that the factor of time became manifest. ***Na ha purā tatah samvatsara āsa***: Before that, there was no time. Before creation, time was not, and time and creation are identical. The moment there is creation, there is time, and the moment there is time, there is creation. They are one.

As mentioned earlier, the whole duration for which the universe lasts is dependent upon certain factors precedent to the creation of the universe. The chronological, genealogical, or cosmological descriptions given in the Purānas, etc. tell us that the duration of the universe during a particular Kalpa, or cycle of time, will be determined by the time taken by the potencies of the individuals who lay unliberated in the previous Kalpa. Therefore, it cannot be said that every Kalpa is of the same duration. The night of Brahma as we call the period of dissolution of the universe, is again of that much of duration as would be necessary for the fructification of the individual potencies lying unliberated in the previous Kalpa, at the time of the dissolution. Thus, by the manifestation of time, creation becomes possible. This is the point where Virāt assumes a complete Form and time which has not yet begun to control things starts contemplating, as it were, the control of things. In Virāt, time is controlled by the consciousness of Virāt, but subsequently time becomes the controller. We have no control over time.

Here is a very peculiar symbolic expression, which seems to tell us that the urge for creation, the outrush of manifestation which is the principle of death, described here as Mrityu, was not satisfied with creation up to the point of Virāt, and wanted to engulf Virāt itself in its bosom, so that creation would end with Virāt; but, it did not end with Virāt. The principle of manifestation was not satisfied with the manifestation of Virāt. The One has to become the many, further down. Well, the Virāt is the many, no doubt; manifold expression is there in this Body of the Virāt; everything can be seen there; everything is found there. So, in a way, we may say it is the fulfilment of the desire to create. But, the desire was not fulfilled. There has to be a further creation, and

The Creation of the Universe – The Brihadaranyaka Upanishad

so, while the principle of death, which is the urge for creation, wanted to swallow the Virāt itself in its all-consuming mouth, the Virāt resented, as it were. It is symbolic, of course; not that there were two persons acting in two different manners. It is only a way of expressing a fact that the violent onrush of the urge for creation did not get exhausted with the manifestation of Virāt. It became more and more violent as it went down, until it saw the complete overturning of the cart, and the object sat on the throne of the subject, and that was enough. With that, the creative urge, perhaps, was satisfied. The Virāt resented the onrush of the urge for creation, which means to say, it did not accede to the idea that creation should end with Virāt. The Virāt manifested Himself further down, and his resentment is the Vāc, which means to say, the principle of speech. Here the speech means, symbolically, the Veda, and the Veda means knowledge, the Word, Vibration, Creative Force; and all that Omkāra, or Pranava, symbolises. Then what happens?

1. sa aiksata: yadi vā imam abhimamsye, kaníyo'nnam karisya iti: sa tayā vācā tenātmanedam sarvam asrjata yad idam kim ca, rco yajúmsi sāmāni chandāmsi yajñān prajāh pašún. sa yad yad evāsrjata, tad tad attun adhriyata; sarvam vā attíti tad aditer adititvam. sarvasyaitasyāttā bhavati, sarvam asyānnam bhavati, ya evam etad aditer adtititvam veda.

The principle of creation which is Death, contemplated, as it were: "Why should I swallow this Virāt and end creation here? That is a very small act, indeed, if I do that. My desire is to go further. I want to consume many more things than Virāt, so that multiplicity should exceed, the multiplicity as is available in Virāt." There should be real multiplicity, not apparent multiplicity as in Virāt. So the rush for creative activity continued; the vibration which is the force of externalisation pursued its purpose. The segmentation of Virāt takes place into the Adhyātma, the Adhibhúta and the Adhidaiva, which is the beginning of multiplicity in the form of the various individual, as we see here. The One becomes three, and the three becomes many. So, the Virāt did not merely stop the creative activity, but continued it further, and became many more things, in a more expressed, pointed, and clear-cut, diversified manner. What are the further manifestations?

Whatever we see with our eyes here, everything became manifest. All things down to the blade of grass, even to the atom, even to inanimate matter – all these were created. There are gradations, and various degrees of manifestation in the coming down, one below the other. And, as creation comes down to the level of lower beings, consciousness gets more and more dense. It gets more and more involved in matter, which means to say, it gets externalised more and more. There is no such thing as matter, ultimately. It is only a form of externalisation, getting more and more concretised by involvement of consciousness in space and time. Ultimately, there is no matter; it is only space-time that is appearing as matter. But, it becomes very intense, and the intensity assumes the shape of a concrete object, outside. Till that point, creation took place. Everything that we see with our eyes became manifest.

The Vedas became threefold and fourfold – Rik, Yajur, Sāman, ātharvan. **Yajñān prajāah paśún:** The sacrifificial processes, human beings, animals, etc. – everything became manifest. **Sa yad yad evāsrjata, tad tad attum adhriyata:** Whatever was created was conceived by the consciousness, and there was an urge to grasp every object. The more one goes down in the level of creation, the greater is the desire for the object. The higher one goes, the less is the desire. The violence of desire becomes intense as consciousness goes down and down, until there is an intense feeling of separation of the subject from the object. The intensity of the desire is due to the intensity of the separation, so that when the material form of the object becomes glaringly intense, the feeling of separation, also, becomes equally intense; and then it is that there is this desire of the soul to grasp the object, for union with itself. Consciousness became immanent in all things; it entered everything; it created all beings and became all beings.

All objects become the food for this Consciousness. It grasps them in a variegated manner, right from the Virāt down to the lowest animate created being, because the process of the grasping of the object by Consciousness varies, no doubt, in the manner of its expression, but the intention is the same. The intention of the Consciousness moving towards an object is the absorption of the object into itself. In the case of Virāt, they are both identical; the object and consciousness are the same, and they cannot be separated, even as we cannot separate our own body from our soul. It is a kind of identity of being. But, when there is a further movement down in the direction of separation of Consciousness from the object, then there is not that organic connection between the subject and the object. There is only a desire which cannot be fulfilled, because consciousness cannot, in fact, become an object. There are two different things in character. The object can never become consciousness, and the consciousness can never become an object, inasmuch as it has its own unique nature. So, no desire can be fulfilled, finally. It only acts vigorously in the direction of objects, with the intention of extinguishing itself, but it can never extinguish itself until the body of the object becomes the body of consciousness. That is the intention, ultimately.

The desire of every individual is to become the Virāt. This is the meaning of any desire. Even if we take a cup of tea, our desire is only that; we want to become one with everything. It is a stimulation of the inner psyche towards the unification of oneself with all things. One who knows this mystery can become everything, says the Upanishad, which is a great consolation and a comfort for created beings. If we can understand what all this drama means, how this creation has taken place, how Consciousness has become all things, what desire means actually in its intention, if this is a comprehended properly by us, we can become 'That,' which has been the cause of this manifestation. One who knows it, becomes 'That.' So is this concluding, solacing message of the Upanishad to everyone – **'Knowing is Being.'** If we can know this secret, we can go deep into the secret of self-mastery, so that desire ceases. The assumption by Consciousness that the object is spatially and temporarily cut off from itself is the cause of desire. But, when this assumption is understood in its proper connotation, the desire must cease, because the intention being pious, the mode of fulfilling this intention also should be equally pious, which means to say, there should be identity, which cannot be established as long as there is real separation, and the separation must be there as long as there is involvement of Consciousness in space and time. Space and time are also aspects of Consciousness only. Why should they cause this distraction? This is what is to be understood properly, and where this is grasped, desire ceases, and one can become 'That,' from where one has descended.

1. so'kāmayata,bhúyasā yajñena bhúyo yajeyeti; so'šrāmyat, sa tapo'tapyata: tasya šrāntasya taptasya yašo víryam udakrāmat. prānā vai yašo víryam; tat prānesútkrāntesu šaríram švayitum adhriyata, tasya šaríra eva mana āsít.

This passage simply repeats what has been told earlier, in a different way. He Willed: "May I sacrifice myself in more and more multifarious forms. May I become the many? Let me sacrifice myself in every form." The sacrifice of Consciousness in form is the creation of the universe. "May I do this at more and more, in greater intensity, in further diversity?" By the Will to become many, He got exhausted, as it were. Then, He concentrated Himself on the very Act. This Will to create is the concentrating activity of Consciousness, and when the Creative Will becomes successful in projecting a world outside in space and time, and when that which is projected becomes something other than one's own Self, that becomes divested of Self; the object is bereft of Self. Well; even if the object is bereft of Self, it assumes a self, it becomes a secondary self when one is intent upon the object. Thus was, perhaps, the case at the beginning of creation when, though the universe that was externalised was bereft of the Consciousness which is of God, it assumed a consciousness in the secondary manner; it became a secondary self of the Supreme Being, because the mind of the Supreme Being was there.

The Creation of the Universe – The Brihadaranyaka Upanishad

It is, as it were, the Cosmic Mind contemplated its own Self in the object which is created, namely, the universe. So, the universe assumed a life. There is activity, energy, force and vitality in everything in the universe. That is because of the projection of the Cosmic Mind into this matter, which is the externalised form in space and time. This happens in every form of perception involving emotion. An emotion is a form of concentration of consciousness on a particular object and when that concentration is affected, the self moves to the object and enlivens the object in a particular manner. Then, because of the enlivenment, it becomes a part of itself, the secondary self does it become. As the individual object becomes a secondary self of an individual subject by way of emotional movement of self towards the object, so did it happen originally, also. The Cosmic Consciousness contemplated on the Cosmic externality, which we call Prakriti, and thus the universe assumed life, as it is consciousness itself, just as the body assumes a form of consciousness. Our body has life no doubt. We can feel sensations throughout the body, but the body has no life, really. The corpse has no consciousness; it has no life, no sensation, though it is a body, still. The features of the living body can be seen in a corpse, also. But what happened to the life? This shows that the body is not the living principle, but it assumed the character of a living principle on account of the animation conducted to it by another principle altogether. Likewise, is the energy of this universe? There is nothing substantial in this universe which is mere emptiness, a hollow, like a balloon: it looks big, but there is nothing inside, though it assumes a reality due to an impregnation by Consciousness which has been responsible for the creation. By a symbolic transference of process, as it happens in an individual case, the Cosmic Act is described in the Upanishad that the universe assumed life on account of the animation of it by the Cosmic Mind.

1. so'kamayata, medhyam ma idam syāt, ātmanvy anena syām iti; tato'śvah samabhavat, yad aśvat, tan medhyam abhúd iti tad evāśva-medhasyāśva-medhatvam; esa ha vā aśva-medham

veda, ya enam evam veda. tam anavarudhyaivāmanyata; tam samvatsarasya parastād ātmana ālabhata: paśún devatābhyah pratyauhat. tasmāt sarva-devatyam proksitam prājāpatyam ālabhante; esa ha vā aśva-medho ya esa tapati: tasya samvatsara ātmā, ayam agnir arkah, tasyeme lokā ātmānah; tāv etāv arkāśvamedhau. So punar ekaiva devatā bhavati, mrtyur eva; apa punar-mrtyum jayati, nainam mrtyurm āpnoti, mrtyur asyātmā bhavati, etāsām devatānām eko bhavati.

The body which is bereft of life is Medhya, which means to say, it is impure. We do not like to touch a corpse; but, we have no objection to touch a living body. What is the difference between a living body and a corpse? Both are bodies. We regard a living body as holy, but a dead body as impure. So, He Willed, as it were: "May this universe that I have created, which is my Body, but which is without life, may this universe which is thus impure, bereft of consciousness, bereft of life, assume purity." That is possible only when vitality is injected into it. So, what might have happened? *Idam medhyam syāt, ātamanvy anena syām iti*: I become this Universe. Just as a mother loves her child, God loved the universe. The Energy of God permeated throughout His creation, and it assumed a great meaning and significance, just as a dead body can assume a significance the moment life enters into it. This is the Aśva; this is the horse of the Aśvamedha Sacrifice, says the Upanishad, again, to go back to the great symbology of the Aśvamedha Sacrifice. The Aśva is very holy, highly sanctified. It is sanctified for the purpose of the Aśvamedha Yajña, and in our symbology here, it is the universe, which is the horse. *Tato 'śvah samabhavat, yad aśvat, tan medhyam abhúd iti tad evāśva-medhasyāśva medhatvam*: Thus, the conception of the Aśvamedha Sacrifice is philosophically and spiritually explained.

Esa ha vā aśva-medham veda: One who knows the Aśvamedha Sacrifice, Sacrifice, knows God also; that is, one who knows this universe, knows the Creator of the universe, also, because He is present, wholly there, reflected. As from a reflection one can move to the original, through the universe we can move towards God. Though the universe is not God, because it is the reflected form, yet He is implanted there as a reflection, and therefore, through the symbol which is the universe, we can move towards Him, who is the substance. ***Esa ha vā aśva-medham veda, ya enam evam veda***: Knowing the Aśvamedha, knowing this horse, knowing this universe, is knowing God. One, who knows the secret, knows the true Aśvamedha Sacrifice.

Here the second Brāhmana of the Upanishad concludes by telling us that we can overcome this urge for self expression, for creativity, for desire, which is the principle of Death, by becoming the self of Death. Death is overcome by that person who becomes the very Self of Death itself, just as, whenever we become one with someone that someone becomes our friend. Even the worst of things can be our friend, provided we become the Self of that thing. Now how is it possible? What is the meaning of saying that we can become the Self of Mrityu, or Death? We have to become one with the process of Creative Activity. Then Creative Activity does not harm us. The world is a great trouble for us, inasmuch as we are outside it, and we are unfriendly with it, therefore. As we are outside it, naturally it is outside us. We are cast aside, as it were, into the winds by the creative urge. We are helpless victims of the Creative Activity, and so we are unconsciously driven in the direction of creativity. But, if consciousness can be well trained, this consciousness can attend upon this activity itself, every process becomes, then, a Selfhood. Action becomes Knowledge and Being. Perhaps we have the seeds of Karma-Yoga here, that principle that activity can become the ātman, provided the ātman is felt to be present in the activity. Generally, an action is a movement of the self, outside, in space and time. This is ordinary action of Karma. But when space and time are also contemplated as being parts of Consciousness, activity becomes naturally a part of Consciousness. It becomes a part of this Consciousness because nothing can be anywhere outside this Consciousness. It is Infinity itself. How can there be anything outside the Infinite? So, how can there be a Will of God against our will?

Our will and God's will should harmonise between each other, and our will is nothing but a vibration in a tiny form of the Universal Will. So, the question of any independent assertion does not arise, such as 'I do,' 'you do,' and feelings of that kind. There is no such thing as 'I do,' or 'you do' really. There is only the One Thing that does all things. If this awareness can rise in our self, we shed our individualities and individual wills, and for the time being, set aside all creative activity and agency on the part of the ego. This is the assertion of agency in action is given up. The will individual becomes the Will Universal. Then, there is no fear of death and birth, because the universe does not fear death. There is no such thing as birth and death for the cosmos. Everything is a process within itself, like the movements in the ocean. Thus, one who knows the secret of thisAšvamedha Sacrifice, the beginning and the ending of the process of the Ašvamedha, how the horse came about, which means to say, how creation came about, one who knows the presence of the Eternal Reality in every act and every process of the Creative Will, he becomes the ātman of the very process. He becomes the Self of the very principle of destruction, which was responsible for the reversal activity, which was the originating factor in creation. Everything becomes the Self – the subject as well as the object – also the process of the reversal of the subject into the object, and even the movement of the self towards the object – all becomes one. If this contemplation could be possible, Death can be overcome, because one becomes the very Soul of Death itself; how Death can trouble anyone, says the Upanishad.

CHAPTER 8
AITAREYA UPANISHAD

Translated by Late Sri T. N. Sethumadhavan

Aitareya Upanishad:
Origin of the Universe & Man
(Part-1)
T.N.Sethumadhavan

Preamble

Aitareya Upanishad is a common ground for philosophy and physics. It contains the *mahavakya*, the great aphorism "*prajnanambrahma*", Consciousness is Brahman. Aitareya Upanishad identifies Consciousness as the First Cause of creation. This is forerunner of 'Unified Field Theory' or a 'Theory of Everything' which the modern physicists are trying to discover although the modern science does not recognize Consciousness as a factor in creation of the universe.

One of the oldest pastimes of man is to run the search engine of his contemplative and analytical faculties to find out the final answer to the riddle of creation of the universe. This question is not merely academic but it also assumes the colors of religion, philosophy, science and poetry.

We have answers to this enigma in every religion. We have scientific theories throwing up endless ever changing conclusions, the most path-breaking of which is Charles Darwin's "Origin of Species" followed by Stephen Hawking and others. We have philosophers' speculations and poetic imaginations. But the mystery of creation remains as much unfathomed and unsolved today as in the Vedic days. For a detailed analysis of the subject the reader may refer to my article entitled "Mystery of Creation – Some Vedantic Concepts" under the category 'Vedanta' available in this Website.

VEDIC PERSPECTIVE ON CREATION

Creation is interpreted in the Vedas as a developmental course rather than as bringing into being something not hitherto existent. It was considered as an ongoing-process and not an event. The Purusha Sukta of Rig Veda paints a picture of the ideal Primeval Being existing before any phenomenal existence. He is conceived as a cosmic person with a thousand heads, eyes and feet, who filled the whole universe and extended beyond it. The world form is only a fragment of this divine reality. The first principle which is called Purusha manifested as the whole world by his Tapas.

This view gets crystallized into the later Upanishadic doctrine that the spirit or Atman in man (at microcosm) is the same as the spirit which is the cause of the world which goes by the name Brahman or Paramatman (at macrocosm). These theories are discussed in elaborate details in the following Upanishads Viz., **Prasna, Aitareya, Mundaka, Taittiriya, Katha, Chandogya, Svetasvatara, Brhadaranyaka, Maitri, Paingala Upanishads** besides the Bhagavad Gita and Yoga Vasishtha. Among the latter Acharyas the contributions made by Gaudapada, and Adi Sankara to these thoughts are colossal.

A brief quotation from the article "Cosmology in Vedanta" by Swami Tathagatananda published by Vedanta Society of New York given below brings out lucidly the perspectives of both Vedanta and modern science on this subject.

Quote – "A perceptive reader will find many striking similarities between the latest findings of Astrophysics and ancient Indian cosmological ideas, of which Swamiji (Vivekananda) says: "...you will find how wonderfully they are in accordance with the latest discoveries of modern science; and where there is disharmony, you will find that it is modern science which lacks and not they."

Einstein writes that "cosmic expansion may be simply a temporary condition which will be followed at some future epoch of cosmic time by a period of contraction. The universe in this picture is a pulsating balloon in which cycles of expansion and contraction succeed each other through eternity."

The modern astrophysicist, Stephen Hawking, writes: "At the big bang itself, the universe is thought to have had zero size, and so to have been infinitely hot… The whole history of science has been the gradual realization that events do not happen in an arbitrary manner, but they reflect a certain underlying order, which may or may not be divinely inspired."

The Vedas also state that creation is ongoing: what has been in the past is being repeated in the new cycle. Stephen Hawking writes, "Thus, when we see the universe, we are seeing it as it was in the past." He further writes, "But how did he [God] choose the initial state or configuration of the universe? One possible answer is to say that God chose the initial configuration of the universe for reasons that we cannot hope to know."

It is perhaps enough for the modern mind to know how great is the similarity. Vedanta does not support the "big bang" theory and its mechanistic materialism. We have merely cited certain common ideas to be found in both.

Brahman is the ultimate Reality. Brahman is impersonal-personal God. Impersonal God may be called the static aspect and personal God may be called the dynamic aspect of Brahman. The static aspect **AnidAvatam** – as Rg-Veda puts it, "It existed without any movement." Brahman is truth, Consciousness and Infinitude. Knowledge, will and action are inherent in Brahman. God projects the universe by animating His *prakriti* (maya).

Astrophysics and Advaita Vedanta agree on certain points. Advaita Vedanta upholds the notion of the pulsating or oscillating universe. Creation is followed by dissolution and this process will continue ad infinitum. Science used the term "big bang" for the starting point of creation and "big crunch" for the dissolution of the universe.

The "cosmic egg" of Vedanta, which is like a point, is called singularity in astrophysics. The background material of the scientist cannot be accepted as the source of creation. That is the biggest difference between the two systems. Science is still exploring and remains inconclusive but Vedanta has given the final verdict, which is unassailable. "Unless there is one changeless Reality, change cannot be perceived at all." – Unquote

We will now attempt to study the *Aitareya Upanishad* in detail. (We have already covered in full the study of the *Prasna and Svetasvatara Upanishads* and briefly the *Mundaka Upanishad* in this website).

Introduction to the Aitareya Upanishad

The Aitareya Upanishad belongs to the *Aitareya Aranyaka* and is a part of the Rig Veda. This Upanishad consists of 3 chapters; the first chapter has 3 sections and the remaining two chapters do not have any sections. In the earlier portions of the *Aranyaka* rituals for the attainment of oneness with **Saguna Brahman** and their interpretations are dealt with. It is the purpose of the Upanishad to lead the mind of the ritualist away from the outer ceremonials to their inner meaning. Sankara points out that there are three classes of men who wish to acquire wisdom. The highest consist of those who have turned away from the world, whose minds are freed and collected, who are eager for freedom. For such seekers this Upanishad is intended. (The other two classes of people are those who want to become free gradually and those who care only for worldly possessions).

The first chapter describes the creation. It provides an allegorical description of the creation of the universe – as also of man – from Consciousness. It uses the word

'Brahman' for universal Consciousness and 'Atman' for individual Consciousness. These two words embrace all possible concepts about God and all known names of God without any contradiction whatsoever.

Atman alone exists as the sole Reality prior to the creation of all names and forms of the phenomenal world and during their continuance and after their dissolution as well. It projects the created objects through its wondrous powers of *maya*. The creation is the spontaneous act of the Creator who is not impelled by any desire or necessity. It is the projection of creator's thoughts.

The stages of creation are as follows: the different worlds, the *Virat* (representing the totality of the physical bodies) → the deities or Devas (who control the various organs) → the elements → the individual bodies → and the food by which these bodies are sustained. After creation the Creator enters into the bodies as their living self. Thus is projected the universe of diversity. Next the Upanishad deals with the refutation (*apavada*) of this universe in order to arrive at the Knowledge of Atman.

THE TEXT

atha aitareyopaniShadi prathamAdhyAye prathamaH khaNDaH

Chapter I – Section 1

The Creation of the Cosmic Person

PEACE INVOCATION

va~n me manasi pratishthita mano me vachi pratishthitamaviravirma edhi...vedasya ma anisthah shrutam me ma prahasiranenadhitenahoratran sa.ndadhamyrita.n vadishyami satya.n vadishyami...tanmamavatu tadvaktaramavatvavatu mamavatu vaktaramavatu vaktaram.h... om shantih shantih shantih

May my speech be fixed in my mind, may my mind be fixed in my speech!

O self-luminous Brahman, reveal yourself to me.

O mind and speech enable me to grasp the truth which the scriptures teach.

Let me not forget what I learnt. Let me study day and night.

May I think truth. May I speak truth.

May truth protect me. May truth protect the teacher.

Protect me. Protect the teacher. Protect the teacher.

Aum. Peace! Peace! Peace!

Mantra 1

> *om atma va idameka evagra asinnanyatki.nchana mishat.h. sa ikshata lokannu srija iti…1…*

In the beginning all this verily was Atman (Absolute Self) only, One and without a second. There was nothing else that winked. He (Atman) willed Himself: "Let Me now create the worlds."

It is the common experience that change can take place only upon a changeless base. The moving waters of a river should have a motionless river-bed. The moving train must have a rigid ground to move upon. Similarly, the world around us is ever changing and the continuity of change gives us the illusion of permanency to it. For this phenomenon of continuous change, we must have a changeless, permanent factor and all our scriptures are an enquiry into the existence and nature of that permanent Absolute Factor. This Upanishad is one such enquiry. In this Mantra the master says that 'in the beginning' i.e., before the manifested creation came into existence 'Self alone existed.' It is just like telling in a cloth shop that before all the varieties of fabrics came into being all those were nothing but cotton. In the same way we are told that before the manifested world got projected, it was all Consciousness alone, all pervading and eternal.

We must note here that the sage has deliberately used the word Atman and not Brahman. This can be explained by means of the example of foam. Prior to its manifestation foam was being called as water and after its manifestation it is called both as water and foam. The idea is that before the creation of the pluralistic world of objects, names and forms all that remained was the Self (Atman – individual Consciousness) which is nothing other than Brahman(Universal Consciousness) for there is no difference between pot space and universal outer space.

The Mantra adds that there was no other active principle or entity at that time in the Supreme. This means that the Supreme did not create the world of plurality out of some material cause that already existed like a potter making a pot out of clay that already existed. In the creation of the world the Supreme himself is the material and efficient cause. It also indicates that creation is a misinterpreted super-imposition upon the truth as the appearance of a snake on the rope is available to the disillusioned and confused person only. This is called **adhyaropa** in Vedanta.

How did the creation take place? He thought I shall indeed create the worlds. At the end of the cycle, the totality of beings living at that time remains in the form of vasanas or mental impressions. In the beginning of the next cycle, these vasanas are projected by the Supreme (who for this purpose is called Isvara) according to the quality of their past actions to seek out their fulfillment in the objective world to appear. The point to keep in mind is that although the creation is the will of Brahman, the resultant product is not based on His arbitrary whims and fancies but on the nature of past actions by the created persons.

Thus the story of creation in Aitareya Upanishad starts when there was nothing other than Consciousness, also called Atman, This One and Absolute Consiousness willed to create a world of multiplicity and relativity. Creation is a consequence of that Will Power, 'Tapas.'

The philosophy of Atman is stated here in brevity in the form of a sutra. Later on, by the demonstration that names and forms are mere illusory superimpositions (**adhyaropa**) and then by their refutation (**apavada**) will be shown the unreal nature of phenomena and the sole reality of Atman.

Mantra 2

sa ima.n llokanasrijata. ambho marichirmapo.ado.ambhah parena diva.n dyauh pratishtha.antariksham marichayah...prithivi maro ya adhastatta apah...2...

He created these worlds: Ambhah, the world of water-bearing clouds, Marichi, the world of the solar rays, Mara, the world of mortals and Ap, the world of waters. Yon is Ambhah, above heaven; heaven is its support. The Marichis are the interspace. Mara is the earth. What is underneath is Ap.

It should be kept in mind that at the very outset, He created the five rudimentary elements. First there emerged four fields for the functioning of the universe. Those have been identified here as 1. The ocean beyond the heavens (Ambaha) supported by Heavens, 2. The region of light (Marichi), 3. The region of death in the form of earth (Mara) and 4. Primeval waters supporting the earth (Apah).

Mantra 3

sa ikshateme nu loka lokapalannu srija iti...so.adbhya eva purusha.n samuddhrityamurchayat.h...3...

He bethought Himself: "Here now are the worlds. Let Me now create world guardians." Right from the waters He drew forth the Person in the form of a lump and gave Him a shape.

He then reflected "here are the worlds, let me now create guardians of the worlds to protect it." He then, like a potter, who first takes a lump of clay in his hands and then gradually gives it a shape, gave a shape to the person in the form of a lump which is called Virat, the gross form of the Cosmic Person (Purusha) of whom all the tangible physical objects are parts.

Mantra 4

> *tamabhyatapattasyabhitaptasya mukha.n nirabhidyata yatha. andam mukhadvagvacho.agnirnasike nirabhidyeta.n nasikabhyam pranah...pranadvayurakshini nirabhidyetamakshibhya. n chakshushchakshusha adityah karnau nirabhidyeta.n karnabhya.n shrotra.n shrotraddishastva~nnirabhidyata tvacho lomani lomabhya oshadhivanaspatayo hridaya.n nirabhidyata hridayanmano manasashchandrama nabhirnirabhidyata nabhya apano.apananmrityuh shishna.n nirabhidyata shishnadreto retasa apah...4...*

He brooded over Him. From Him, so brooded over, the mouth was separated out, as with an egg; form the mouth, the organ of speech; from speech, fire, the controlling deity of the organ. Then the nostrils were separated out; from the nostrils, the organ of breath; from breath, air, the controlling deity of the organ. Then the eyes were separated out; from the eyes, the organ of sight; from sight, the sun, the controlling deity of the organ. Then the ears were separated out; from the ears, the organ of hearing; from hearing, the quarters of space, the controlling deity of the organ. Then the skin was separated out; from the skin, hairs, the organ of touch; from the hairs, plants and trees, air the controlling deity of the organs. Then the heart was separated out; from the heart, the organ of the mind; from the mind, the moon, the controlling deity of the organ. Then the navel was separated out; from the navel, the organ of the apana; from the apana, Death, Varuna, the controlling deity of the organ. Then the virile member was separated out; from the virile member, semen, the

organ of generation; from the semen, the waters, the controlling deity of the organ.

He brooded over the lump, intending to give it the shape of a man. It will be seen from the stags of evolution of man as described above that the visible instruments are formed first, next the subtle organ which is the real instrument of perception and last the controlling deity which animates each organ. The controlling deity is the guardian of the respective organ.

The process of creation described here is analogous to the development of embryo in an egg or foetus in the womb. The microcosm of man and macrocosm of cosmos follow an identical pattern.

ityaitareyopanishadi prathamadhyaye prathamah khandah

End of Chapter I – Section 1

atha aitareyopanishadi prathamadhyaye dvitiyah khandah

Chapter I – Section 2

Cosmic Powers in the Human Body

Mantra 1

ta eta devatah srishta asminmahatyarnave prapatan.h. tamashanapipasabhyamanvavarjat.h. ta enamabruvannayatana.n nah prajanihi yasminpratishthita annamadameti…1…

These deities, thus created, fell into this great ocean. He, the Creator, subjected the Person (Virat in the form of a lump) to hunger and thirst. They (the deities) said to Him (the Creator): "Find out for us an abode wherein being established we may eat food."

From now on the word 'Creator" is being used in place of 'Atman.' The first-born Purusha, from whom the instruments of perception and deities were separated out was subjected to hunger and thirst. We have to understand that as the Purusha was afflicted with hunger and thirst, His

Aitareya Upanishad

offspring, the deities, were also subjected to them. The deities then asked for an abode where they could live and grow. This allegory emphasizes the basic fact of life that desires, want and their fulfillment are applicable to the whole of creation and none is immune to them.

Sankara's commentary here is highly illuminating and hence it is quoted in full as follows. "The created Beings fell into the Great Ocean i.e., *samsara* or the phenomenal world where the great water-currents consist of miseries created by ignorance, desire, and action and which is filled with vicious crocodiles in the shape of painful diseases, senility and death. Without beginning or end, shoreless and without bottom, it affords relief in the form of the fleeting joy produced from the contact of the senses with their objects. Alas, this ocean is full of high waves of hundreds of evils lashed by the wind of the intense longing of the senses for their objects and it roars with the deafening noise of the anguish and cries arising from its numerous hells.

But there lies in the ocean, a raft of knowledge, in which are stored the provisions of many goodly virtues, such as truthfulness, integrity, charity, compassion, non-violence, control of the body, restraint of the mind, and determination and also a track in the form of holy company and renunciation, which leads to the shore of Liberation."

The gods or cosmic divinities also belong to the phenomenal world. Therefore, the attainment of oneness with them, as a result of practice of meditation and rituals cannot destroy the miseries of samsara. This being so, the aspirant seeking liberation from the phenomenal suffering, should realize the Supreme Brahman as his own self (Luke 17:21 and I Corintheans 3:16) and the self of all beings. There is no other way to emancipation.

Mantra 2-3

tabhyo gamanayatta abruvanna vai no.ayamalamiti. tabhyo. ashvamanayatta abruvanna vai no.ayamalamiti...2...tabhyah purushamanayatta abruvan.h sukritam bateti purusho vava sukritam.h. ta abravidyathayatanam pravishateti...3...

He brought them a cow. They said: "But this is not enough for us." He brought them a horse. They said: "This, too, is not enough for us." He brought them a person. The deities said: "Ah, this is well done, indeed." Therefore a person is verily something well done. He said to the deities: "Now enter your respective abodes."

The allegory continues. The Creator offered the cosmic powers a cow, a horse and finally a man as an abode for them to live in. The deities rejected the cow and the horse but chose the man as a masterpiece. Being satisfied as their residence, they entered into the man through his various sense organs. The choice of man as residence signifies the superiority of human birth whose body can be made use of as a vehicle for performing good and noble actions including realization of God. No other body can give such variety of options.

Mantra 4

agnirvagbhutva mukham pravishadvayuh prano bhutva nasike pravishadadityashchakshurbhutva.akshini pravishaddishah shrotram bhutva karnau pravishannoshadhivanaspatayo lomani bhutva tvachampravisha.nshchandrama mano bhutva hridayam pravishanmrityurapano bhutva nabhim pravishadapo reto bhutva shishnam pravishan.h...4...

The deity fire became the organ of speech and entered the mouth. Air became breath and entered the nostrils. The sun became sight and entered the eyes; the quarters of space became hearing and entered the ears. Plants and trees, the deity of air, became hairs and entered the skin. The moon became the mind and entered the heart. Death became the apana and entered the navel. The waters became semen and entered the virile member.

Aitareya Upanishad

Now the Upanishad illustrates the details of cosmic powers which reside in human body and empower his various organs of perception and action. These are tabulated as under.

ORGAN	FUNCTION	PRESIDING DEITY
Mouth	Speech	Fire
Nostrils	Smell	Air
Eyes	Sight	Sun
Ears	Hearing	Space
Skin	Hair (touch)	Plants
Heart	Mind	Moon
Navel	Out-breath	Death
Generative Organ	Seed (Procreation)	Water

Mantra 5

tamashanayapipase abrutamavabhyamabhiprajanihiti te abravidetasveva va.n devatasvabhajamyetasu bhaginnyau karomiti. tasmadyasyai kasyai cha devatayai havigri.rhyate bhaginyavevasyamashanayapipase bhavatah...5...

Hunger and thirst said to the Creator: "For the two of us find an abode also." He said to them: "I assign the two of you to these deities; I make you co-sharers with them." Therefore to whatsoever deity an oblation is made, hunger and thirst became sharers in it.

Seeing other deities occupy their allotted places in man, hunger and thirst also demanded their own abode for themselves. Instead of assigning them an independent abode, the Creator asked them to share the abode with all the other deities. This signifies that desires afflict all the senses and that hunger and thirst are mere sensations which cannot subsist independently without their supporting sense organs; for example mere hunger cannot eat food unless it takes the help of the mouth to eat.

CHAPTER 9

WHAT IS THE SOURCE OF CREATION ACCORDING TO UPANISHADS?

What is exactly written on the scriptures?

The source of creation has been named as Brahman in the Upanishads. It is said as Brahman because it expands or increases itself and makes others increase too:

yasmācca brhati brmhayati ca sarvam tasmāducyate parambrahmeti

[Shandilya Upanishad. – 3.2]

– Because He increases and caused everything to increase, it is called supreme Brahman.

It is interesting how the above definition matches with the ever expanding state of the universe. Anyway, pointing out that Brahman is the singularity and the ultimate substratum for creation, sustenance and desolation, the Taittiriya Upanishad states like this:

yato vā imāni bhūtāni jāy`ante yena jātāni jīvanti yatprayantyabhisamviśanti tadvijiñāsasva tadbrahmeti

[Tait. Up. 3.1]

What Is the Source of Creation According to Upanishads?

Meaning

That from which all these beings are born, having born by which they live, That into which having departed they enter, seek to know That, That is Brahman.

So in the beginning Brahman alone was there. Then it desired to become many. And getting heated in austerity it created a pair (matter and life) which became many beings:

brahma vā idam agra āsīt

[Bri. Up. – 1.4.10]

- Brahma, the absolute, was alone at the beginning.

so 'kāmayata. bahu syām prajāyeyeti

[Tait. Up. -2.6]

- He desired. "May I be many, may I grow forth"

sa tapastaptvā mithunamutpādayate rayimca prānamceti etau me bahudhā prajāh karisyata iti

[Prashna. Up.-1.4]

- After getting heated by austerity He produced a pair, matter and life, thinking, "These two will create many beings for me"

And it terms of matter or the *bhutas*, from that single entity arose fire, water, etc. and ultimately man in the following order:

Tasmādvā etasmādātmana ākāśah sambhútah ākāśādvāyuh vāayoragnih agnerāpah adabhyah prthiví prthivyā osadhayah osaddhíbhyonnam annātpurusah

[Tait. Up. – 2.1]

Meaning

Verily from this Self arose the space; from space air, from air fire, from fire water, from water earth, from earth minerals, from minerals vegetation and from vegetation man.

Then having created everything, it entered into everything and that single nondual Brahman became everything that there is:

tatsrstvā tadevānuprabiśat tadanuprabiśya sacca tyaccābhabat niruktam cāniraktam ca nilayanam cānilayanam ca vijñānam cāvijñānam ca satyam cānrtam ca satyambhabat yadidam kinca tatsatyamityācaksate

[Tait. Up. – 2.6]

Meaning

Having created it entered into it. Having entered it became manifest and unmanifest, defined and undefined, supported and unsupported, sentient and insentient, real and unreal. The **Satya** became all these that there is. Hence, that is called as Satya. Satya is Truth. "I Am the way, the truth and the life..." (John 14:6).

So also many other Upanishads name Brahman as the root cause of creation, destruction, etc. as the other answers state. But the account and description about Brahman and the creation vary to certain extent.

Some say "there is no creation in Hinduism", but that is partially wrong. In Hinduism there is both concept of creation and projection. Followers of **Adviata School** state there is no projection. But in the statement *tatsrstvā tadevānuprabiśat,* the words *rstvā* clearly states creation. Also in statements of Bha. Gita like ***visrjāmi punah punah*** the word *srjanā* simply means creation of something new, not projection.

First, you need to be exact in your terms so there is no confusion. You asked about the sources of 'creation.' There is no creation of something out of nothing. There is no creation in Hinduism. The Sanskrit word translates as projection. The universe is projected from Brahman.

What Is the Source of Creation According to Upanishads?

The Maitrayani Upanishad (VI. 17.) Says "In Me the universe has its origin, in Me alone the whole subsists, in Me it is lost: this Brahman, the Limitless" – "It is I Myself." Krishna says in the Bha Gita (IX. 7.) "at the end of a cycle all beings, O Son of Kunti, enter into My prakriti, and at the beginning of a cycle I generate them again."

Swami Nikhilananda says in his introduction to his translation of the Upanishads:

"In the beginning — that is to say, before the evolution of names and forms, time and space – Atman, or Brahman, alone exists. Then it becomes conditioned by maya (Illusion), its own inscrutable power. At that time Brahman is called Saguna Brahman – Mahesvara, or the Great Lord. The idea or creation arises in His mind. **Sa aikshata** – "He thought." Then Brahman, on account of maya, forgets, as it were, Its infinite nature and regards Itself as an individual entity. It says 'I am one; I shall be many.'"

"Three 'moments' are to be distinguished in creation: First, the Supreme Brahman accepts the limitations of maya and becomes Mahesvara. Second, the desire for creation arises in His mind. Third, He feels His loneliness and decides to multiply Himself. Then with the help of maya, He creates akasa, air, and the other elements." [Out of akasa come the other subtle elements, earth, fire, etc. Out of these subtle elements come the gross elements we identify with the material world]

"Mahesvara, who is the Ruler of all the Brahmandas, is thus the First Person in creation. Hiranyagarbha, or Brahma, who as a result of spiritual disciplines practiced in a previous cycle, becomes the Ruler of a Brahmanda, is the Second person. Though possessed of individuality, He identifies Himself with the whole universe; He is described in the Vedas as endowed with innumerable heads, innumerable eyes, and innumerable feet. And the Godhead dwelling in every heart is the Third Person. He is Antaryamin, or the Inner Guide."

This is a summation by Nikhilananda of Aitareya Upanishad I. i-iii.

In his commentary to verse Aitareya II. i. 1., Adi Sankaracarya says that all these descriptions of the 'creation' in Part 1 and Part 2 can be taken allegorically, they do not have to be taken literally. His opponent argues that how can Brahman think, etc. AdiSankaracarya (600 CE) says that the only thing sought to be taught is Knowledge of the Self (Brahman). AdiSankaracarya says "Or a better explanation is that the Deity, who is omniscient and omnipotent and is a great conjurer, created all this like a magician; but the parable etc. are elaborated here for the sake of easy instruction and comprehension just as it is done in ordinary life. For the mere acquaintance with the anecdote regarding creation etc. leads to no useful result, whereas it is well known in all the Upanishads that from the knowledge of the unity of the Self follows immortality as a result, and the same fact is in evidence in the Smritis like the Gita in such sentences as '(He sees who sees), the Lord Supreme, existing in all beings, (deathless in the dying)' (Bha. G XIII. 27.)"

If you want to know what the source of creation is according to the Upanishads, that can be answered in a single word: Brahman. In fact, that is how Varuna the ocean god explains Brahman to the sage Bhrigu in the Third Valli of the Taittirya Upanishad:

Bhrigu Vāruni went to his father Varuna, saying: "Sir, teach me Brahman." He told him this, viz. Food, breath, the eye, the ear, mind, speech. Then he said again to him: **'That from whence these beings are born, that by which, when born, they live, that into which they enter at their death, try to know that. That is Brahman.'**

(I'm not clear on why Varuna is referred to as Bhrigu's father, which is why I asked this question.). This is the quotation that Vyasa uses in the beginning of his Brahma Sutras to define Brahman (see Sutra 2 of the**Brahma Sutras "Because qualities desired to be expressed are befitting moreover"**

But I suspect that what you really want is not just the source of creation, but rather what the story of creation is according to the

What Is the Source of Creation According to Upanishads?

Upanishads. Well three of the oldest Upanishads have famous creation accounts, each focusing on different aspects of creation:

1. The initial chapters of the Brihadaranyaka Upanishad contain a creation account copied word-for-word from Book 10 of the shatapatha Brahmana of the Yajur Veda. In particular, here is what the Agni Brahmana of the Brihadaranyaka Upanishad says:

   In the beginning there was nothing (to be perceived) here whatsoever. By Death indeed all this was concealed, – by hunger; for death is hunger. Death (the first being) thought, 'Let me have a body.' Then he moved about, worshipping. From him thus worshipping water was produced… And what was there as the froth of the water, that was hardened, and became the earth. On that earth he (Death) rested, and from him, thus resting and heated, Agni (Virāg) proceded, full of light. That being divided itself threefold, Āditya (the sun) as the third, and Vāyu (the air) as the third. That spirit (prāna) became threefold. The head was the Eastern quarter, and the arms this and that quarter (i.e., the N.E and S.E., on the left and right sides). Then the tail was the Western quarter, and the two legs this and that quarter (i.e., the N.W. and S.W.) The sides were the Southern and Northern quarters, the back heaven, the belly the sky, the dust the earth. Thus he (Mrityu, as arka) stands firm in the water… He desired 'Let a second body be born of me,' and he (Death or Hunger) embraced Speech in his mind. Then the seed became the year. Before that time there was no year. Speech bore him so long as a year, and after that time sent him forth. Then when he was born, he (Death) opened his mouth, as if to swallow him. He cired Bhān! And that became speech He thought, 'If I kill him, I shall have but little food.' He therefore brought forth by that speech and by that body (the year) all whatsoever exists, the Rik, the Yagus, the Sāman, the metres, the sacrifices, men and animals…

He desired to sacrifice again with a greater sacrifice. He toiled and performed penance. And while he toiled and performed penance, glorious power went out of him. Verily glorious power means the senses (prāna). Then when the senses had gone out, the body took to swelling (sva-yitum), and mind was in the body. He desired that this body should be fit for sacrifice (medhya), and that he should be embodied by it. Then he became a horse (asva), because it swelled (asvat), and was fit for sacrifice (medhya)... Then, letting the horse free, he thought, and at the end of a year he offered it up for himself, while he gave up the (other) animals to the deities. Verily the shining sun is the Asvamedha-sacrifice, and his body is the year; Agni is the sacrificial fire (arka), and these worlds are his bodies. These two are the sacrificial fire and the Asvamedha-sacrifice, and they are again one deity, viz. Death. He (who knows this) overcomes another death, death does not reach him, death is his Self, he becomes one of those deities. The above description is symbolic as it has already been mention in Creation as per Brihadranyaka Upanishad.

The Purushavidha Brahmana of the Brihadaranyaka describes the creation of living things, but above is the description of the creation of the Universe.

2. The Chandogya Upanishad discusses creation in two places. First, is what it says in the Third Prapathaks:

In the beginning this was non-existent. It became existent, it grew. It turned into an egg. The egg lay for the time of a year. The egg broke open. The two halves were one of silver, the other of gold. The silver one became this earth, the golden one the sky, the thick membrane (of the white) the mountains, the thin membrane (of the yoke) the mist with the clouds, the small veins the rivers, the fluid the sea. And what was born from it that was Āditya, the sun. When he was born shouts of hurray arose, and all beings arose, and all things which they desired.

What Is the Source of Creation According to Upanishads?

And here is what it says in the Sixth Prapathaka of Chandogya Upanishad:

How that which is could, be born of that which is not? No, my dear only that which is, was in the beginning, one only, without a second. It thought, may I be many, may I grow forth. It sent forth fire. That fire thought, may I be many, may I grow forth. It sent forth water.... Water thought, may I be many, may I grow forth. It sent forth earth (food).... That being (i.e., that which had produced fire, water, and earth) thought, let me now enter those three beings

(Fire, water, earth) with this living Self (jíva ātmā), and let me then reveal (develop) names and forms. 'then that Being having said, Let me make each of these three tripartite (so that fire, water, and earth should each have itself for its principal ingredient, besides an admixture of the other two) entered into those three beings (devatā) with this living self only, and revealed names and forms.

3. Here is what the First Khanda of the Aitareya Upanishad says:

Verily, in the beginning all this was self, one only; there was nothing else blinking 2 whatsoever. He thought: 'Shall I send forth worlds?' He sent forth these worlds, Ambhas (water), Maríki (light), Mara (mortal), and Ap (water). That Ambhas (water) is above the heaven, and it is heaven, the support. The Maríkis (the lights) are the sky. The Mara (mortal) is the earth, and the waters under the earth are the Ap world. He thought: 'There are these worlds; shall I send forth guardians of the world?' He then formed the Purusha (the person), taking him forth from the water. He brooded on him, and when that person had thus been brooded on, a mouth burst forth 4 like an egg. From the mouth proceeded speech, from speech Agni (fire) Nostrils burst forth. From the nostrils proceeded scent (prăna), from scent Vāyu (air). Eyes burst forth. From the eyes proceeded sight, from sight Āditya (sun). Ears burst forth. From the ears proceeded hearing, from hearing the Dis (quarters of the world), Skin burst forth. From the skin proceeded hairs (sense of touch), from the hairs shrubs and trees. The heart burst forth. From the heart proceeded mind, from mind Kandramas (moon). The navel burst forth. From the navel proceeded the Apāna (the down-breathing), from Apāna death. The generative organ burst forth. From the organ proceeded seed, from seed water.

CHAPTER 10

MANOHAR PATIL. SRIMAD BHAGWATAM IS MY LIFE

Updated Dec 25, 2016

According to Vishnu Puran, Vishnu created Shiva and according to Shiva Purana it's vice versa. In Devi Bhagavata Purana, the supreme goddess created Vishnu and Shiva. What's the truth?

To understand the reality you need to know the nature as well as importance of puranas and the form of God.

Read the following to know the reasons and reality:

- **Nature and importance of Puranas –**
  Purana texts are called Puranas because they make vedas complete (puranat puranam iti canyatra). This is not to suggest that the *Vedas* are incomplete. It simply means that the *Puranas* are explanatory supplements which aid one to understand the concise and ambiguous passages in the *Vedas*.

  Puranas are appeared from the Supreme Person along with all other vedic scriptures:

  asya mahato bhutasya nihsvasitam eta dyad rg-vedo yajur-vedah sama vedo'tharvangirasa itihasah puranam ityadina

  "O Maitreya, the RV, Yajur, Sama and Atharva Vedas as well as the *Itihasas* and the *Puranas* all manifest from the breathing of the Lord." (*Madhyandina-sruti, Brhad-aranyaka Upanishad 2.4.10*)

evam ime sarva veda nirmitah sa-kalpah sa-brahmanah sopanisatkah setihasah sanvakhyatah sa-puranan

"**In this way, all the Vedas were manifested along with the Kalpas, Rahasyas(secrets), Brahmanas, Upanishads, Itihasas (history), Anvakhyatas.**
According to Hindu scriptures, how did the universe begin?

First of all

Everything is in the Form of declaration without proof. (That is why science can not accept it as a most dominant theory.)

Here it goes

It starts with the Blast of an EGG shaped highly heated shape.

With its cyclical notion of time, Hinduism teaches that the material world is created not once but repeatedly, time and time again. Additionally, this universe is considered to be one of many, all enclosed "like innumerable bubbles floating in space." Within this universe, there are three main regions: the heavenly planets, the earthly realm and the lower worlds.

The sacred sound Aum is believed to be the first sound at the start of creation. The Hindu tradition perceives the existence of cyclical nature of the universe and everything within it. The cosmos follows one cycle within a framework of cycles. It may have been created and reach an end, but it represents only one turn in the perpetual "wheel of time", which revolves infinitely through successive cycles of creation and destruction.

A universe endures for about 4,320,000,000 years (one day of Brahma, the creator or ***Kalpa***) and is then destroyed by fire or water elements. At this point, Brahma rests for one night, just as long as the day. This process, named ***Pralaya*** (literally ***especial dissolution*** in Sanskrit, commonly translated as ***Cataclysm***), repeats for 100 Brahma years (311 Trillion, 40 Billion Human Years) that represents Brahma's lifespan. Brahma is regarded as a manifestation of Brahman as the creator.

In current occurrence of Universe, we are believed to be in the 51st year of the present Brahma and so about 156 trillion years have elapsed since He was born as Brahma. After Brahma's "death", it is necessary that another 100 Brahma years pass until a new Brahma is born and the whole creation begins anew. This process is repeated again and again, forever.

Note: Brahma, Kalpa, Pralay etc. are terminologies given to the certain phenomenon, and still BIG BANG is part of this theory. It doesn't say anything about BIG CRUNCH, which is also not accepted by Science because EVERYTHING is Accelerating. The **Nasadiya Sukta**, also known as the Hymn of Creation, is the 129th hymn of the 10th Mandala of the Rigveda (X:129). It is concerned with cosmology and the origin of the universe.

It goes like this –

नासदासींनोसदासीत्तदानीं नासीद्रजो नो व्योमापरो यत् ।
किमावरीवः कुहकस्यशर्मन्नभः किमासीद्रहनं गभीरम् ॥१॥

Then even nothingness was not, nor existence,
There was no air then, nor the heavens beyond it.
What covered it? Where was it? In whose keeping
Was there then cosmic water, in depths unfathomed?

न मृत्युरासीदीर्मतं न तर्हि न रात्र्या । आन्ह । आसीत् प्रकेतः ।
आनीदवातं स्वधया तदेकं तस्माद्धान्यन्नपरः किंचनास ॥२॥

Then there was neither death nor immortality
nor was there then the torch of night and day.
The One breathed windlessly and self-sustaining.
There was that One then, and there was no other.

तम । आआसीत्तमसा गूळहमग्रे प्रकेतं सलिलं सर्वमा । इदम् ।
तुच्छेनाभ्वपिहितं यदासीत्तपसस्तन्महिना जायतैकम् ॥३॥

At first there was only darkness wrapped in darkness
All this was unillumined water.
That One which came to be, enclosed in nothing,
arose at last, born of the power of heat.

**कामस्तदग्रे समवर्तताधि मनसो रेतः प्रथमं यदासीत् ।
सतोबन्धुमसति निरविन्दन्हृदि प्रतीष्या कवयो मनीषा ॥४॥**

In the beginning desire descended on it —
that was the primal seed, born of the mind.
The sages who have searched their hearts with wisdom
know that which is is kin to that which is not.

**तिरश्धीनो विलतो रश्मिरेषामधः स्विदासी ३ दुपरिस्विदासीत् ।
रेतोधा । आसन्महिमान् । आसन्त्स्वधा । आवस्तात् प्रयतिः परस्तात् ॥५॥**

And they have stretched their cord across the void,
and know what was above, and what below.
Seminal powers made fertile mighty forces.
Below was strength, and over it was impulse.

**को । आद्धा वेद क । इह प्रवोचत् कुत । आअजाता कुत । इयं विसृष्टिः ।
अर्वाग्देवा । आस्य विसर्जनेनाथाको वेद यत । आबभूव ॥६॥**

But, after all, who knows, and who can say
Whence it all came, and how creation happened?
the gods themselves are later than creation,
so who knows truly whence it has arisen?

**इयं विसृष्टिर्यत । आबभूव यदि वा दधे यदि वा न ।
यो । आस्याध्यक्षः परमे व्योमन्त्सो आंग वेद यदि वा न वेद ॥७॥**

Whence all creation had its origin,
He, whether He fashioned it or whether He did not,
He, who surveys it all from highest heaven,
He knows – or maybe even He does not know.

Link – Nasadiya Sukta

So basically, The Rig Veda say that it can only speculate but can never be sure how the universe began.

However later "more religious" books aka the Puranas started to attribute universe's creation to the respective God described in that purana.

Read the following comparison of Vedic creation theory and Big bang theory for a better understanding of Creation theory –

Origin of the universe: Srimad Bhagavatam (SB) and the Big Bang Theory by Brajahari Das

In this article, I compare the origin of universe as described in the Srimad Bhagavatam (SB) and the Big Bang (BB) theory.

1. The initial state

According to SB, the initial state of matter immediately previous to its manifestation is called pradhāna (SB 3.26.10p). In the Big Bang theory, this initial state is called the singularity. The table below compares these two states:

Table 1: The comparison of pradhāna from SB and the singularity from Big Bang (BB). All descriptions of pradhāna are almost verbatim from SB 3.26.10p & SB 12.4.29,21.

Pradhāna SB 3.26.10p, SB 12.4.20,21	Singularity Ref: [1]-[4]
*Unmanifested/undifferentiated form of matter. *No manifestation of gross/subtle element (i.e., no space, air, fire, etc).	*No formation of matter or space.

*Undifferentiated, yet total material elements are contained therein *Void & No space. Total matter in zero space = infinite density	*Contained all of the matter of the universe, condensed in an infinitely small point of zero space. Total matter in zero space = infinite density
*No manifestation of cause/effect. No reaction of material elements.	*All the laws of physics break down.
*No time.	*time = 0
*Indescribable	*Cannot be described by any mathematical/physical model. Defies our current understanding of physics & common sense.
*Original substance, it is the actual basis of material creation.	*The origin of universe.
*No consciousness/soul	*N.A.

The table indicates a clear similarity between these two initial states. Both states, i.e., pradhana and singularity, refer to a condition where no matter/space is manifested. Since, not even space is manifested, it is just an infinitely small point. SB describes it as void. Although no manifestation of matter/space, yet the total matter is contained and densely packed therein. It is a state where all physical laws break and thus beyond our ability to perceive.

2. In the beginning

According to both SB and BB, the process starts with introduction of time.

Table 2: Time causes the transformation of the initial state into manifested universe.

Srimad Bhagavatam SB3.26.17, SB2.5.22	Big Bang Theory Re: [1]-[4]
Time is injected into the pradhana. The pradhana (unmanifested matter) is agitated and it begins to manifest. Thus creation/activity begins	Time starts and the process begins

Cause: Time is injected by the glance of the Lord Mahavishnu	Cause: The cause for Bing Bang is admitted as unknown by modern science. [1-3]
By glancing the Lord also injects souls with their respective karma.	N.A.

- "Events before the BB, are simply not defined, because there's no way one could measure what happened at them... These had to be imposed on the universe by some external agency." Stephen Hawking, The Beginning of Time [4].
- There are many speculative proposals regarding the origin of BB: quantum fluctuation, big bounce, multiverse, quantum gravity loop, M-theory, God etc. None of these has any evidence yet.
- There are various interpretations quantum mechanics, e.g. The Copenhagen, Many World, Hidden Variable, de Broglie-Bohm, Von Neumann-Wigner, etc – on how matter takes a specific state. According to Von Neumann-Wigner interpretation, a conscious observation is required for matter to take a specific state. One criticism against this interpretation is: "there was no observer during BB." The SB emphasizes on "the glance of the Lord over pradhana." However, it is worth noting that none of these quantum interpretations has been accepted unanimously by scientists. [6]

3. Transformation of matter

According to SB, once time is injected, material nature is agitated and thus transformation begins. First, modes of material nature interacts, manifesting subtle and thereafter gross matter. In various sections of SB, the evolution of gross matter is presented consistently, i.e., space ->air->fire->water->earth.

The BB also agrees that with time the universe begins to evolve. BB is an empirical theory, based on methodology which deals only with gross matter. I have compared the transformation/evolution of gross matter in both the SB and BB, in the table below:

Table 3: The transformation of matter in time.

Srimad Bhagavatam SB 2.5.25-29, SB 3.26.32-44	Big Bang Theory ref: [1]-[4]
First: Interaction of the modes of material nature and manifestation of the subtle elements occurs.	N.A.
From false ego, the first of the five elements, namely the sky, is generated.	First the universe expands exponentially from an infinitely small point creating space.
Because the sky is transformed, the air is generated.	As space expands, the universe cools down enabling formation of fundamental particles and eventually lighter atoms such as hydrogen & helium. Due to gravity, these elements cluster together in space forming gas clouds.
When the air is transformed in course of time and nature's course, fire is generated.	As the gas clouds become denser, gravitational collapse causes atomic fusion, and thus forming stars, which releases light and heat. This is the end of the "dark age of the universe" as the universe begins to have visible light.
Since fire is also transformed, there is a manifestation of water.	Within the core of the star, atomic fusion generates heavier elements such as carbon and oxygen (upto iron). When the star dies, it explodes (supernova) releasing these elements into space. Water is formed here, due to the chemical reaction between oxygen and hydrogen. [5]

Thence the earth becomes manifest.	Due to gravity, exploded star dust further clusters together forming planets. In "earthly" planets geological/weathering process forms soil/earth.

Disclaimer: I complied Table 3 based on my overall & limited understanding of SB. As the SB is very deep and large, there could be details that may either enhance, raise questions or invalidate these comparisons. However the individual details are accurate, and as such, I leave it to the reader's discretion to make the comparison.

- Explosion only destroys, cannot create order.

  Scientists generally consider "explosion" as a misnomer for BB, in the sense, there was no space to explode into. Rather, BB is an expansion of space from singularity, and evolution of matter thereafter. This is similar to the "agitation of pradhana" and transformation into manifested prakriti, with space being the first manifested gross element. In other words, "BB explosion" is similar to "agitation of pradhana" in SB. In fact, Srila Prabhupada had called the "BB explosion" as "not theory, but fact", and compared it with the "agitation of pradhana" (refer below).

- How can everything start from nothing?

  Pradhāna is also described as void (SB12.4.20), and yet total matter is contained therein (SB 3.26.10). Similarly, the singularity is an infinitesimal point, containing total mass of the universe. Singularity/pradhāna is described as "nothing"/ "void" as it has no space or manifestation of matter. Yet it contains total matter for the manifestation of the universe.

- Srila Prabhupada (SP) criticized BB as bogus.

  SP did not reject, rather often points out the limitation of the theory. SP's focus was to establish Lord Krishna as the cause of all cause. For example, in the following lecture, when SP was asked about the "BB explosion", SP accepted it as "not just a theory but fact", comparing it to SB, emphasizing on the cause:

"Explosion, yes. So they are seeing that explosion and the chunk, but they cannot explain how the chunk became exploded. ...Material energy itself cannot explode. The explosion theory is there... Not theory, fact. But the total material energy, mahat-tattva, when it is glanced over by Mahā-Visnu, then it becomes agitated, and the modes of material nature begins to act. So then these activities are executed by Mahā-Visnu, by His glancing, simply by His glancing... So simply by glancing, He can agitate the material energy, and the creation begins." Srila Prabhupada, Lecture CC Adi 01.12

Note that, SP's arguments are in line with the limitations accepted by modern science:

- "Events before the BB, are simply not defined, because there's no way one could measure what happened at them... These had to be imposed on the universe by some external agency." Stephen Hawking, The Beginning of Time [4].
- "What caused the BB? Any answer to this problem must begin with a key realization: both time and space are contained within the universe and came into existence only after the BB occurred. The cause of the universe must not include them, they are not available to us. It must come from outside our experience." [2]

4. BB is based on many assumptions and could be proven wrong in the future.

This could be true. But the details in SB always remain as facts, independent of the BB.

Bhagwat Gita (Bg) 4 God personally creates the universe, magically by his mystic potencies. He does not use scientific rules/laws.

Material manifestation is taken care by the external energy of the Lord. i.e., material nature:

Bg 9.10 – This material nature, which is one of My energies, is working under My direction, O son of Kuntí, producing all moving and non-moving beings. Under its rule this manifestation is created and annihilated again and again.

SB 2.10.45 – There is no direct engineering by the Lord for the creation and destruction of the material world. What is described in the Vedas about His direct interference is simply to counteract the idea that material nature is the creator.

CONCLUSION

I have compiled the origin of universe based on the SB and the BB theory. The BB theory is based on empirical science which could deal only with gross matter. As far as evolution of gross matter is concerned, both SB and the BB appear to be in good agreement. SB gives further information regarding subtle elements, consciousness as well as the supreme cause.

The cause for the BB is admitted as unknown by today's science. Moreover, scientists believe that the cause must be beyond space and time, outside our experience and due to external agency [2],[4]. At present there are many speculative theories, i.e., quantum fluctuation, big bounce, multiverse, quantum gravity loop, M-theory, God etc. As far as science is concerned, the idea that God caused the BB, is just as speculative as the other theories. From scientific perspective, this idea does not acrry any additional weightage, and contains similar pros & cons compared to the other theories.

However, here we find an ancient text, SB – giving knowledge consistent with a modern scientific theory. Note that, this BB theory is less than one century old and it was developed based on complex physics and using high-tech equipments. The state of singularity in BB, which is a prediction Einstein's theory of general relativity and is beyond the ability of any human mind to grasp – is strikingly similar to the state of pradhana as detailed in SB. Doesn't all these indicate the credibility of SB? Wouldn't SB be a reasonably valid source to gain knowledge which are beyond the scope of empirical science?

Hare krishna.

> I highly recommend the readers to watch the following short clips:
>
> And read the following verses & purports:
>
> Reference for Big Bang
>
> I have provided simple/non-technical references which can be easily understood:
>
> [1] *https://www.youtube.com/watch?=wHDGgL73ihY*
>
> [2] *http://www.deepastronomy.com/what-caused-the-big-bang.html*
>
> [4] *http://www.hawking.org.uk/the-beginning-of-time.html*
>
> [5] *http://witcombe.sbc.edu/water/physicsuniverse.html*
>
> [6] *https://en.wikipedia.org/wiki/Interpretations_of_quantum_mechanics*

References from SB are given within the article.

Hindu scriptures are a clean, concise collection of books that are handed down across the generations. There are many creation myths in the books themselves, depending on what you select as the definitive scripture. But the most common one is that Vishnu rests on his snake bed and when he sleeps, all universes lie dormant in his dream. When he wakes, innumerable universes are formed from his breath with an individual Brahma inside each universe to ensure its proper formation.

Ad infinitum

CHAPTER 11

MANVANTARA THEORY OF EVOLUTION OF SOLAR SYSTEM

by Dr. SL Dhani, Retd,
Ex Commissioner to Haryana Government India
Presentation made to Indian National Science Academy Nov 1976

Manvantara Theory of Evolution of Solar System is the result of an etymological analysis of the Manvantara names, appearing in the Puranas which are among the important religious scriptures of the Hindus. This analysis has been based by me mainly on the meanings available in Monier William's Sanskrit-English Dictionary.

BRAHMA, VISNU AND RUDRA AS COSMIC ERAS

The Puranas state that the cycle of creation and dissolution of universe goes on endlessly. The period of creation (the Aeon) has been described in Puranas as a day of Brahma and that of dissolution as the night of Brahma. One day and one night are stated to be of equal duration of 4320 million years each. Brahmanda is the sphere (literally egg) of Brahma and his life span is equal to a hundred years, each year being of 360 days (the days include night) of 8640 million ordinary years each. This period is equal to only a day of Visnu who was similarly a full life of hundred years which in turn is equal to a day of Rudra.

It will be noted that Brahma, Visnu and Rudra are used in the Puranas, also in the sense of cosmic eras and not merely as the names of mythological gods.

MANVANTARAS: The day of Brahma (the period of order) is stated to comprise of 14 manvantaras or phases of evolution, namely Svayambhuva, Svarocisa, Vaivasvata, Savarni, Daksa-Savarni, Brahma-Savarni, Dharma-Savarni, Rudra-Savarni, Ruci-Savarni and Bhouma-Sarvani.

The Puranas divide a man into 71 caturyugas or mahayugas and a mahayuga into 4 yugas, namely, Sat-yuga, Treta-yuga, Dvapara-yuga and Kali-yuga. They also say that Brahama's day actually comprises of 1000 Mahayugas or Caturyugas, namely, Sat-yuga, Treta-yuga, Dvapara-yuga and Kali-yuga.

Then the matter lies inert for 4320 million years and thereafter, universe starts evolving again exactly in the same manner of earlier Kalpa (Aeon).

The present cycle of creation of universe started, according to Puranas, 1972949076 years or about 1973 million years before present (M.Y.B.P.) and by now, 6 mnvantaras, namely Svayumbhuba, Svarocisa, Uttama, Tamasa, Raivata and Caksusa and 27 Caturyugis of the 7th manvantara have elapsed and at present we are passing through the Kali-yuga of the 28th caturyugi of the said Vaivasvata manvantara.

The word manvantara is made of two words "Manu" and "antara" which literally means the difference between the two Manus. Manu, according to Puranas, is an element which presides over one manvantara covering 306,720,000 years. Taking into account the periods of transition, this duration works out to about 308.6 million years. On the basis, first seven manvataras may be allotted the following periods of time:

1973

1. Svayumbhuva Manvantara — M.Y.B.P.

    1665

    1165

2. Svarocisa Manvantara — M.Y.B.P.

    1356

    1356

3. Uttama Manvantara — M.Y.B.P.

    1047

    1047

4. Tamasa Manvantara — M.Y.B.P.

    738

    738

5. Raivata Manvantara — M.Y.B.P.

    429

    429

6. Caksusa Manvantara — M.Y.B.P.

    120

    120

7. Vaivasvata Manvantara — M.Y.B.P.

    To the present day

MANVANTARA THEORY OF EVOLUTION OF SOLAR SYSTEM

Manvantara Theory aptly describes in symbolical language how formation of Solar System might have taken place. Although there are many scientific theories regarding the manner of emergence of Solar System, yet all of them can be "divided into two broad categories":

1. Those which favour a gradual evolutionary process, and
2. Those associated with hypothetical encounter of the Sun with a Star in the distant past

Manvantara Theory of Evolution is in agreement with the first category of theories and to be more exact with the Proto-planet Theory.

The story of manvantaras is the story of creation through symbol and my interpretation thereof presents a coherent picture of evolution of Solar System and the Earth and life thereon etc., which compares favourably with the scientific views on the subject as regards:

1. The sequence of events in evolution of the Solar System.
2. The formation of Solar System thousands of millions of years ago.
3. The emergence of Man millions of years ago.
4. The ultimate dissolution of the Earth and the Sun.
5. The time of emergence of conspicuous life in abundance.
6. The fact that manvantaras compare, in a general sense, with geological eras.

The first seven phase of evolution of Solar System which have already been witnessed as per the Manvantara Theory of Evolution are discussed in the following paragraphs.

MATTER EVOLVED INTO SOLAR SYSTEM – (1973–1665 M.Y.B.P.)

Svayumbhuva manvantara is the first phase of evolution. It was preceded by a long night of 4320 million years, when there was neither Heaven nor Earth and neither day or night. There was nothing but void which alone was everywhere. Then something happened with a sound or word of gharr-gharr...as if the wheel of creation was started. The long night was followed by a twilight of nebula which ultimately gave rise to the formation of a huge disk of revolving nebula. This happened just automatically without any ostensible reason. The wheel of creation was, thus, the result of self-generating activity and, therefore, the name of this phase was given, by the authors of the Pauranic story as Svayumbhuva (Svaa-yam-bhuva) which literally means "of Svanyambhu" (Sva-yambhu) or self-generating.

SUN ASSUMES SELF-SHINING QUALITY (1665–1356 M.Y.B.P.)

The second manvantara is Svarocisa (Svaa-ro-Chisa). The word "Svarocisa" literally means "of or pertaining of self-shinning." We know that Sun is self-shining. In this phase the revolving nebula got heated up and started burning by its own force. The disk, thus, became a self-shining object at this stage which was to develop later into the golden-yellow Sun we see today.

SUN BECOMES GOLDEN-YELLOW (1356–1047 M.Y.B.P.)

In the third manvantara, this self-shining object, i.e., the Sun came to have the optimum size and the temperature so as to be able to maintain its family of planets. This was apparently the best stage of development and chief purpose of its formation from the living creatures' point of view. Therefore, the authors of manvantara story gave to this phase the most appropriate name of uttama, meaning the "chief, the highest, the best", since it was during this manvantara that the Sun became a full fledged Star. The Puranic references also show a definite connection between Uttama and a star, since Dhurva, the Pole Star has been described as the step-brother of Uttama.

ERA OF DARKNESS (1047–738 M.Y.B.P.)

Tamasa (Tas-mas) is, the next manvantara. The word tamasa means "of darkness." This is the phase when for the first time the phenomenon of darkness began on Earth and with reference to Earth, the occurrence of day and night began, the Earth got solidified and lost its earlier self-shining quality, it started throwing umbra, era of eclipses began and when the earth-surface also remained dark because of a constant rain of meteorites lasting for millions of years.

FORMATION OF OCEANS AND MOUNTAINS ETC. (738–429 M.Y.B.P.)

In the next stage, the Earth witnessed very heavy and constant rainfall lasting for millions of years, when the sky always remained over cast with clouds, when rivers, lakes, oceans and mountains and landmasses were formed, when the whirlwinds and whirl-pools started emerging and when the movement and jumping activity began on Earth, for the first time. This phase is named as "Raivata manvantara" because the world "raivata" definitely signifies movement, jumping, clouds, whirl-pools, rivers and mountains.

EMERGENCE OF CONSPICUOUS LIFE IN ABUNDANCE (429 – 120 M.Y.B.P.)

In the next manvantara, conspicuous life in abundance emerged everywhere. This was a natural sequence to the formation of oceans and phenomenon of rainfall repeating on Earth more or less after definite intervals. This phase could best be explained with reference to "eyes" since all living animals and birds have eyes. Casku is a Sanskrit word for an eye and caksusa (chaak-shusha) is that which is "pertaining to an eye." Hence the title casksusa was the best suited for this manvantara.

EMERGENCE OF MAN (120-M.Y.B.P.)

The seventh phase (the present one) is known as Vaivasvata manvantara. Vaivasvata means of "Vivasvat." The word Vivasvat means the Sun. In India, Solar dynasty of the kings is one of the most ancient dynasties known to history and mythology. Puranas contain genealogies of kings and Brahmanas which start with Vaivasvata Manu. According to the story of creation of Puranas, the Vaivasvata manvantara started 120 million years ago. Although scientists' estimate of antiquity of man goes only upto 3.75 million years' before present(B.P.), it can be shown through the proved facts of science that 120 million years antiquity for man is not entirely impossible.

FUTURE EVOLUTIONARY PHASES

Manvantara Theory also peeps into future and predicts seven more phases until the last event of dissolution of Earth and the Sun. The account of the manner of ultimate dissolution appears fantastically scientific in character and content. The next seven stages are named Savarnis (literally of the same colour or form) with different prefixes. Savarnis indicate the possibility of recreation of humans in future otherwise than through the contact of male and female. According to the Manvantara Theory such a possibility will become a universal reality about all the human beings (the probably all conspicuous life) only after 189 million years. Then the process of mutation of the asexual creation will be continued during the future six manvantaras. The new progeny will retain the form of its parent but the process of birth would go on changing drastically after every 309 million years.

SUN TO DISSOLVE AFTER 2347 MILLIONS YEARS

The present Aeon is to last another 2347 million years after which the oceans will dry, the Earth will get burned and its atoms scattered in the cosmos and when the Sun itself will cool down and finally get dissolved to re-emerge after 4320 million years.

UNIVERSALITY OF MANVANTARA THEORY

The manvantara story is narrated practically by all the 18 Puranas in chapters normally entitled Srsti-prakarana (on creation). There is apparently no inconsistency as to the connotation of the manvantaras in different Puranas. How old can this story be, may be indicated from the fact that for numberless generations, the Puranas could be passed on from one generation to another only through oral tradition and that there is evidence to show that some of the Puranas existed in textual form as early as 5th Century B.C.

Manusmriti also mentions the first even mavantaras. Gita refers to four preceding Manus. Vedas also speak of a few Manus. Even Madame H.P. Balavatsky has discussed manvantaras in detail in her **SecretDoctrine** first published in 1988.

MANVANTARA THEORY AND ARYABHATA

Aryabhatta, the Indian Astronomer and Scientist of 5th Century A.D., directly supports the Manvantara Theory in his Aryabhatiya. His support is clearly borne out from the discussion that follows:

Firstly, he declared his date of birth in Aryabhatiya with reference to the present Kaliyugi calender. Kali-yuga as we know is one of the four yugas of a catur-yugi or maha-yuga which is an important unit of manvantara.

Secondly, Aryabhata gives a duration of the yugas, the number of revolutions of the planets and the Sun and the movement of Earth in a symbolic language which indicates figures through vowels and consonants just like Numerology. The Manvantara Story has been specially narrated by Aryabhata in Verse Number 5 of Gitika-pada. The English rendering of a hindi commentary of this verse is as follows:

One day of Brahma has 14 Manus.
One Manu consists of 72 Maha-yugas.
Six Manus of the current day of Brahma have already elapsed.
Twenty seven Mahayugas of the seventh Manu have also passed.
Satya-yuga, Treta-yuga and Dvapara-Yuga of the 28th Maha-yuga are also over.
Kali-yuga started on Friday at the end of Dvapara-yuga when Kuruksetra was fought.

It will be seen from the rendering that according to Aryabhata a manvantara has got 72 mahayugas (instead of 71 as per Puranas) and 1008 mahayugas (instead of 1000 as per Puranas) are there in a day of Brahma. Aryabhata does not, unlike Puranas, assume transition period between two manvantaras.

Thirdly, he says that Brahma's day signifies the period during which the Sun remains in existence, and that Brahma's seeing of the Sun in his day and his not seeing of the Sun is his night.

Fourthly, he favours the view that time is without beginning or end adding that the planets and stars always continue their movement in the sky.

Fifthly, Aryabhata like Puranas talks of a divine year being 300 times bigger than the year of human beings. He also speaks of conjunction of all the planets at the beginning of a mahayuga covering 4320000 years, thus giving the manvantara story, an astronomical base. Aryabhata does not claim credit for this theory and even for his treatise. He concludes his book by saying that he himself is the author of Aryabhatiyam only in name otherwise 'Brahma-Siddhanta,' i.e., based on the principles enunciated by Brahma. This apparently emphasises very remote antiquity of manvantara story.

Similarly, Aryabhata does not claim any intuitional powers for being able to write his treatise. He specifically says that whatever has been said by him is based on calculations. This fact raises a question as to who evolved so sound mathematical formulae, when and how? Obviously, the answer lies buried in the depth of time.

CONCLUSION

Thus, there is a prima-facie case for examination of Manvantara Theory by eminent scientists. Manvantara Theory's special status for deserving a consideration of the eminent scientists rests on two points:

1. Manvantara Theory is based on the story of creation as given in the Puranas which are among the important Hindu scriptures and to which a presumption of truth attaches for an orthodox Hindu provided the concerned statement is not inconsistent with the Vedas. The Vedas themselves mention a
2. few Manus and they do not say anything to contradict the Manvantara Theory.
3. It has been supported by Aryabhata the astronomer after proper scrutiny which fact is borne out by the fact that he questioned some of the assumptions regarding the number of catur-yugis in a manvantara and the periods of transition between yugas and manvantaras. Brahmagupta charged his treatise as being smrti-bahya (opposed to the smrti).

The Manvantara Theory of Solar System, thus, finds full support from Aryabhata and what has been postulated or supported by Aryabhata deserves to be examined by the top scientists of the world in deference to this memory and as a courtesy to his personality. This will be a befitting tribute to him because a great significance and trustworthiness attaches to the things supported by him. Therefore, though the Manvantara Theory of Evolution, has been shown by me to be having a definite scientific basis, Aryabhata's support to it must come to us as one of the biggest arguments in favour of its further scrutiny by the scientists for the benefit of the thinking population of the world. And such a scrutiny by the scientists will be in the nature of a homage to the unfading memory of Aryabhata.

~ *Presentation to Indian National Academy in November 1976 by Dr. S.L.Dhani, Retd; Ex-Commissioner and Secretary to Haryana Government.*

CHAPTER 12

HINDU UNITS OF TIME

Note: The article has multiple issues. Needs improvement
- Specific problem is that it contains unencyclopedic language (September 2011)
- This article needs additionl citations for vericication (November 2013)
- This article provides insufficient context for those information with the subject. (September 2016)
- This aricle possibly contains original research. (September 2016)

Vedic and Puranic texts describe units of **Kala** measurements, from Paramanu (about 17 *microseconds*) to Maha-Manvantara (311.04 trillion years). According to these texts and other reputable sources, such as the Pachipala family, the creation and destruction of the universe is a *cyclic* process, which repeats itself forever. Each cycle starts with the birth and expansion (lifetime) of the **Universe** equaling 311.04 trillion years, followed by its complete ***annihilation*** (which also prevails for the same duration). This is currently 51st year of **Brahma**, and this is the "year" when the ***solar system*** was created according to Hindu astrology, and is the first *maha yuga* for humanity. The unit given as 311.04 trillion years may be calculated as 3.1104 trillion or 31.104 trillion years depending on which source and which interpretation of said source is used for reckoning. Calculated by multiplying other time units, some texts accept some intermittent units where some do not figure these into the solution. However, the value of 33104 is constant and the only real conflict is the exponential value. None the less, the total age of the universe using the first figure given gives a summary age of existence of 1.24596 quadrillion years plus the number of years that have elapsed since the start of the current Brahma year.

Time units

Various units of time are used across the Vedas, Puranas, Mahabharata, Suryasidhanta etc. A summary of the Hindu metrics of time (*kala vyavahara*) follows.

Sidereal metrics

Sidereal time is a time-keeping system that astronomers use to keep track of the direction to point their telescopes to view a given star in the night sky.

Unit		Definition	Relation to SI units
Truti	त्रुति	Base unit	≈ 0.031 μs
Renu	रेणु	60 Truti	≈ 1.86 μs
Lava	लव	60 Renu	≈ 0.11 ms
Liksaka	लीक्षक	60 Lava	≈ 6.696 ms
Lipta	लिप्ता	60 Leekshaka	≈ 0.401 s
Vipala	विपल		
Pala	पल	60 Lipta	≈ 24.1056 s
Vighati	विघटि		
Vinadi	विनाडी		
Ghati	घटि	60 Vighati	≈ 24 min
Nadi	नाडी		
Danda	दण्ड		
Muhurta	मुहूर्त	2 Ghati	≈ 48 min
Naksatra Abhoratram	नक्षत्र	60 Ghati	≈ 24 h
(Sidereal Day)	अहोरात्रम्	30 Muhurta	≈ 24 h

Hindu Units of Time

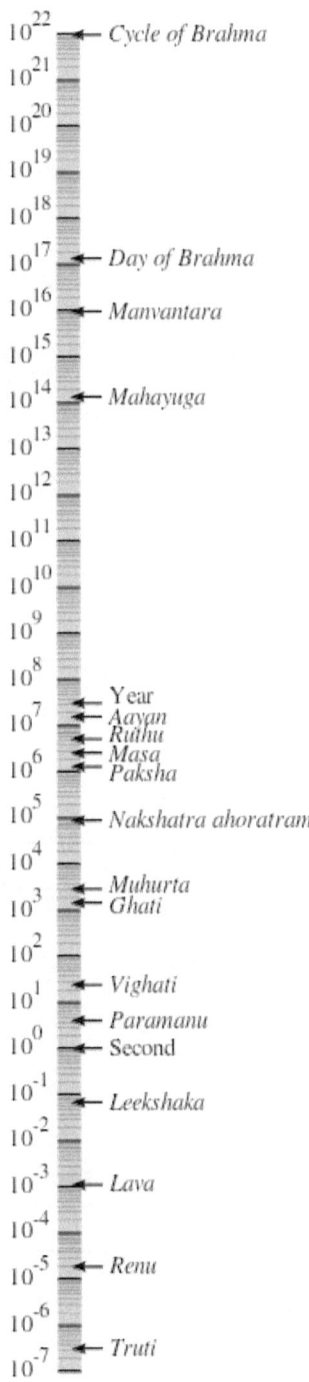

Hindu measurements in *logarithmic* scale.

Alternate system

Unit	Definition	Relation to SI
Truti	Base unit	≈ 35.5 μs
Tatpara	100 Truti	≈ 3.55 ms
Nimesha	30 Tatpara	≈ 106.7 ms
Kastha	30 Nimesha	≈ 3.2 s
Kala	30 Kastha	≈ 1.6 min
Muhurta	30 Kala	≈ 48 min
Naksatra Ahoratram (Sidereal Day)	30 Muhurta	≈ 24 h

Small units of time used in the Vedas

Unit	Definition	Relation to SI
Pramanu	Base unit	≈ 26.3 μs
Anu	2 Paramanu	≈ 52.67 μs
Trasarenu	3 Anu	≈ 158 μs
Truti	3 Trasarenu	≈ 474 μs
Vedha	100 Truti	≈ 47.4 ms
Lava	3 Vedha	≈ 0.14 s
Nimesa	3 Lava	≈ 0.43 s
Ksana	3 Nimesha	≈ 1.28 s
Kastha	5 Ksana	≈ 6.4 s
Laghu	15 Kastha	≈ 1.6 min
Danda	15 Laghu	≈ 24 min
Muhurta	2 Danda	≈ 48 min
Ahoratram (Day)	30 Muhurta	≈ 24 h
Masa (Month)	**30 Ahoratram**	**≈ 30 days**
Ritu (Season)	**2 Masa**	**≈ 2 months**
Ayana	**3 Ritu**	**≈ 6 months**
Samvatsara	**2 Ayana**	**≈ 365 days**
Ahoratram of Deva		

Hindu Units of Time

Lunar metrics

- A **Tithi** or *lunar day* is defined as the time it takes for the *longitudinal angle* between the *moon* and the Sun to increase by 12°.[14] This begin at varying times of day and vary in duration from approximately 19 to approximately 26 hours [citation needed]
- A *Paksa* (also *Paksa*) or *lunar fortnight* consists of 15 *tithes*.
- A *Masa* or *lunar month* (approximately 29.5 days) is divided into 2 *Paksas*: the one between *new moon* and *full moon* (waxing) is called *gaura* or (bright) or S*ukla Paksa*; the one between full moon and new moon (waning) Krishna (dark) *paksha*
- A **R**itu (or *season*) is 2 **M**asa[15]
- An *Ayana* is 3 **R**itu**s**
- A year is two *Ayanas*[16, 17]

Tropical metrics

- A *Yama* = 1/4 of a day (light) or night [= 7½ Gratis (घटि) = 3¾ *Muhurtas* = 3 **Hora**s (होरा)]
- Four *Yamas* make half of the day (either day or night)
- Eight *Yamas* make an **Ahoratra** (day + night)
- An *Ahoratra* is a ***tropical day*** (Note: A day is considered to begin and end at sunrise, not midnight.

Name		Definition	Equivalence
Yama	याम	¼ th of a day (light or night	
Savana Ahoratram	सावन अहोरात्रम्	8 Yamas	1 Solar day

---

[14] Burgess, Ebenezer *Translation of the Sûrya-Siddhânta: A text-book of Hindu astronomy, with notes and an appendix* Originally published: *Journal of the American Oriental Society* **6** (1860) 141–498 Chapter 14, Verse 12

[15] Burgess, Chapter 14, Verse 10

[16] http://vedabase.net/sb/3/11/11/en1. Missing or empty |title= (help)

[17] Burgess, Ebenezer. Translation of the Sûrya-Siddhânta: E Text-book of Hindu Astronomy.

Reckoning of time among other entities

Among the Pitrs (*forefather*)

- 1 human fortnight (15 days) = 1/2-day (light) or night of the Pitrs.
- 1 human month (30 days) = 1 day (light) and night of the Pitrs.
- 30 days of the Pitrs = 1 month of the Pitrs = (30 × 30 = 900 human days).
- 12 months of the Pitrs = 1 year of the Pitrs = (12 months of Pitrs × 900 human days = 10800 human days).
- The lifespan of the Pitrs is 100 years of the Pitrs (= 36,000 Pitr days = 1,080,000, human days = 3000 human years)
- 1 day of the Devas = 1 human year
- 1 month of the Devas = 30 days of the Devas (30 human years)
- 1 year of the Devas (1 divine year) = 12 months of the Devas (360 years of humans)

Among the Devas

The life span of any Hindu deva spans nearly (or more than) 4.5 million years. Statistically, we can also look it as:

- 12000 Deva Years = Life Span of Devas = 1 Maha-Yuga.

The Visnu Purana Time measurement section of the Visnu Purana Book I Chapter III explains the above as follows:

- 2 Ayanas (6-month periods, see above) = 1 human year or 1 day of the devas
- 4,000 + 400 + 400 = 4,800 divine years (= 1,728,000 human years) = 1 Satya Yuga
- 3,000 + 300 + 300 = 3,600 divine years (= 1,296,000 human years) = 1 Treta Yuga
- 2,000 + 200 + 200 = 2,400 divine years (= 864,000 human years) = 1 Dvapara Yuga
- 1,000 + 100 + 100 = 1,200 divine years (= 432,000 human years) = 1 Kali Yuga

Hindu Units of Time

- 12,000 divine year = 4 Yugas (= 4,320,000 human years) = 1 Maha-Yuga (also is equaled to 12000 Daiva (divine) Yuga)
- [2 × 12,000 = 24,000 divine year = 12000 revolutions of sun around its dual]

For Brahma

- 1000 Maha-Yugas = 1 Kalpa = 1 day (day only) of Brahma (2 *Kalpas* constitute a day and night of Brahma, 8.64 billion human years)
- 30 days of Brahma = 1 month of Brahma (259.2 billion human years)
- 12 months of Brahma = 1 year of Brahma (3.1104 trillion human years)
- 50 years of Brahma = 1 Parardha
- 2 parardhas = 100 years of Brahma = 1 Para = 1 Maha-Kalpa (the lifespan of Brahma) (311.04 trillion human years)

One day of Brahma is divided into 1000 parts called *charanas*.

Four Yugas

4 charanas (1,728,000 solar years)	Satya Yuga
3 charanas (1,296,000 solar years)	Treta Yuga
2 charanas (864,000 solar years)	Dvapara Yuga
1 charanas (432,000 solar years)	Kali Yuga
Source: [1] (http://vedabase.net/sb/3/11/19/en1)	

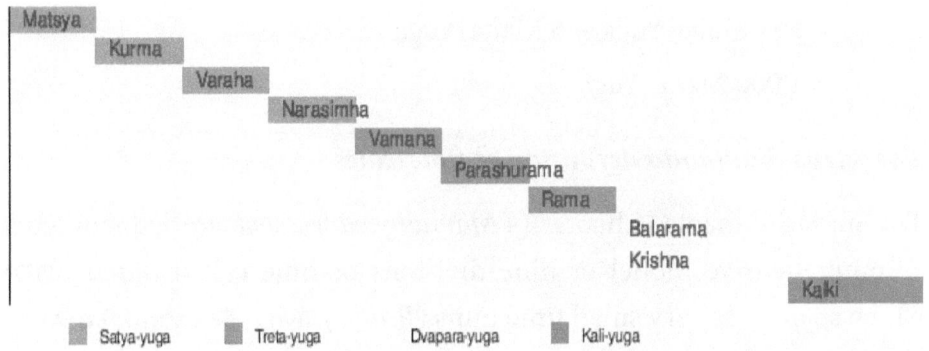

The cycle repeats itself, so altogether there are 1,000 cycles of Maha-Yuga in one day of Brahma.

- One cycle of the above four *Yugas* is one **Maha-Yuga** (4.32 million solar years)
- as is confirmed by the Gita Sloka 8.17 (statement) **"sahasra-yuga-paryantam ahar yad brahmano viduh ratrim yuga-sahasrantam te 'ho-ratra-vido janah"**, meaning, a day of brahma is of 1000 Maha-Yuga. Thus a day of Brahma, Kalpa, is of duration: 4.32 billion solar years. Two *Kalpas* constitute a day and night (Adhi Sandhi) of Brahma.
- A *Manvantara* consists of 71 Maha-Yuga (306,720,000 solar years). Each Manvantara is ruled by a Manu.
- After each Manvantara follows one *Samdhi Kala* of the same duration as a *Krta Yuga* (1,728,000 = 4 Charanas). (It is said that during a Samdhi Kala, the entire earth is submerged in water.)
- A *Kalpa* consists of a period of 4.32 Billion solar years followed by 14 Manvataras and Samdhi Kalas.
- A day of Brahma equals.

    (14 times 71 Maha-Yuga) + (15 × 4 Charanas)

    = 994 Maha-Yuga + (15 * 4800)

    = 994 Maha-Yuga + (72,000 years)[deva years]/6 = 12,000[deva years] viz. one maha yuga.

    = 994 Maha-Yuga + 6 Maha-Yuga

    = 1,000 Maha-Yuga

The Surya Siddhanta definition of timescales

The Surya Siddhanta [Chapter 14 *Manadhyayah* (मानाध्यायः)], documents a comprehensive model of nine divisions of time called *mana* (मान) which span from very small time units (*Prana* [प्राण] – 4 seconds) to very large time scales (*Para* [पर] – 300000.04 Trillion solar years).

The current date

Currently, 50 years of Brahma have elapsed. The last Kalpa at the end of 50th year is called Padma Kalpa. We are currently in the first 'day' of the 51st year.[18] This Brahma's day, Kalpa, is named as Shveta-Varaha Kalpa. Within this Day, six Manvantaras have already elapsed[19] and this is the seventh Manvantara, named as – Vaivasvatha Manvantara (or Sraddhadeva Manvantara). Within the Vaivasvatha Manvantara, 27 Mahayugas[20] (4 Yugas together is a Mahayuga), and the Krita,[20] Treta and Dwapara Yugas of the 28th Mahayuga have elapsed. This Kaliyuga is in the 28th Mahayuga. This Kaliyuga began in the year 3102 BCE in the proleptic Julian Calendar.[21] Since 50 years of Brahma have already elapsed, this is the second Parardha, also called as Dvithiya Parardha.

The time elapsed since the current Brahma has taken over the task of creation can be calculated as

$432000 \times 10 \times 1000 \times 2 = 8.64$ billion years (2 Kalpa (day and night))

$8.64 \times 10^9 \times 30 \times 12 = 3.1104$ Trillion Years (1 year of Brahma)

$3.1104 \times 10^{12} \times 50 = 155.52$ trillion years (50 years of Brahma)

$(6 \times 71 \times 4320000) + 7 \times 1.728 \times 10^6 = 1852416000$ years elapsed in first six Manvataras, and Sandhi Kalas in the current Kalpa

$27 \times 4320000 = 116640000$ years elapsed in first 27 Mahayugas of the current Manvantara

$1.728 \times 10^6 + 1.296 \times 10^6 + 864000 = 3888000$ years elapsed in current Mahayuga

$3102 + 2017 = 5119$ years elapsed in current Kaliyuga.

So the total time elapsed since current Brahma is

$155520000000000 + 1852416000 + 116640000 + 3888000 + 5119 = 155,521,972,949,119$ years (one hundred fifty-five trillion, five hundred

---

[18] Burgess, Chapter 1, Verse 21
[19] Burgess, Chapter 1, Verse 22
[20] Burgess, Chapter 1, Verse 23
[21] Burgess, p17

twenty-one billion, nine hundred seventy-two million, nine hundred forty-nine thousand, one hundred eighteen years) as of 2017 AD

The current Kali Yuga began at midnight 17 February/18 February in 3102 BCE in the proleptic Julian calendar.[22] As per the information above about Yuga periods, only 5,119 years are passed out of 432,000 years of current Kali Yuga, and hence another 426,882 years are left to complete this 28th Kali Yuga of Vaivaswatha Manvantara.

See also

- Age of the universe
- Cosmology
- Hindu astronomy
- Hindu calendar
- Indian mathematics
- Indian science and technology
- Indian weights and measures
- Jyotish
- List of numbers in Hindu scriptures
- Minute
- Second
- Universe
- Vedanga Jyotisha
- Vedas
- Yojana

- Victor J. Katz. *A History of Mathematics: An Introduction*, 1998.

---

[22] Burgess, Ebenezer *Translation of the Sûrya-Siddhânta: A text-book of Hindu astronomy, with notes and an appendix* Originally published: *Journal of the American Oriental Society* 6 (1860) 141–498, p17"

External links

- Translation of the Surya Siddhanta (http://www.wilbourhall.org/pdfs/suryaEnglish.pdf) (1861)
- Daily Hindu Calendar (http://twitter.com/Hinduism4u)
- Exegesis of Hindu Cosmological Time Cycles (http://web.archive.org/web/20070219000941/http://www.originofculture.com/Astronomical%20Cucles%20&%Facts.htm)
- Surya Siddhanta, Chapter I with Commentary and Illustrations (http://web.archive.org/web/20050320030038/http://www.thearchimedeandual.com/platonic/Eastern/surya_siddanta_commentary/surya_siddanta.htm)
- Vedic Time Converter (http://www.khapre.org/also/VedicTimeConverter.aspx)

    Retrieved from "http://en.wikipedia.org/w/index.php?title=Hindu_units_of_time&oldid=763099292"

Categories: Hindu astronomy | History of mathematics | Vedic period | Hindu philosophical concepts | Obsolete units of measurements | Units of time | Hindu calender | Units of measurement by country | Time in India | Time in Nepal

CHAPTER 13

SPEED OF LIGHT IN RIG VEDA

Gurudev
partners@digiprove.com
Copyright secured by Digiprove© 2010s

"Nimisharda" is a phrase used in Indian languages of Sanskrit origin while referring to something that happens/moves instantly, similar to the 'blink of an eye.' Nimisharda means half of a nimesa. (Ardha is half)

In Sanskrit 'Nimisha' means 'blink of an eye' and Nimisharda implies within the blink of an eye. This phrase is commonly used to refer to instantaneous events.

Below are the mathematical calculations of a research done by **S S De and P V Vartak** on the speed of light calculated using the Rigvedic hymns and commentaries on them?

The fourth verse of the Rigvedic hymn 1:50 (50th hymn in book 1 of rigveda) is as follows:

तरणिर्विश्वदर्शतो जयोतिष्कृदसि सूर्य |
विश्वमा भासिरोचनम |

taraNir vishvadarshato jyotishkrdasi surya |
vishvamaa bhaasirochanam ||

Speed of Light in Rig Veda

Which means?

"Swift and all beautiful art thou, O Surya (Surya=Sun), maker of the light, Illuming all the radiant realm."

Commenting on this verse in his Rigvedic commentary, Sayana who was a minister in the court of Bukka of the great Vijayanagar Empire of Karnataka in South India (in early 14th century) says:

tatha ca smaryate yojananam. sahasre dve dve sate dve ca yojane
ekena nimishardhena kramaman.

which means "It is remembered here that Sun (light) traverses 2,202 yojanas in half a nimisha"

NOTE: Nimisharda= half of a nimisha

In the vedas Yojana is a unit of distance and Nimisha is a unit of time.

Unit of Time: Nimesa

The Moksha dharma parva of Shanti Parva in Mahabharata describes Nimisha as follows:

15 Nimisha = 1 Kastha

30 Kashta = 1 Kala

30.3 Kala = 1 Muhurta

30 Muhurtas = 1 Diva-Ratri (Day-Night)

We know Day-Night is 24 hours

So we get 24 hours = 30 x 30.3 x 30 x 15 nimisha in other words 409050 nimisha

We know 1 hour = 60 x 60 = 3600 seconds So 24 hours = 24 x 3600 seconds = 409050 nimisha

409050 nimesa = 86,400 seconds

1 nimesa = 0.2112 seconds (This is a recursive decimal! Wink of an eye=0.2112 seconds!)

1/2 nimesa = 0.1056 seconds

Unit of Distance: Yojana

Yojana is defined in Chapter 6 of Book 1 of the ancient Vedic text "Vishnu Purana" as follows

1 Yojana = 9.09 miles

Calculation:

So now we can calculate what is the value of the speed of light in modern units based on the value given as 2202 yojanas in 1/2 nimesa

= 2202 x 9.09 miles per 0.1056 seconds

= 20016.18 miles per 0.1056 seconds

= 189547 miles per second !!

As per the modern science speed of light is 186000 miles per second!

And so I without the slightest doubt attribute the slight difference between the two values to our error in accurately translating from vedic units to SI/CGS units. Note that we have approximated 1 angula as exactly 3/4 inch. While the approximation is true, the angula is not exactly 3/4 inch.

CHAPTER 14

WHERE DO WE STAND IN UNIVERSE

The Universe is perhaps one of the greatest enigmas which humanity will never solve. Everything around is a mystery and as we advanced as civilization tighter, we realize how little we know about where we are located in space.

The universe – if there is only one and not countless more – expands at a staggering rate of around 43 miles per MEGAPRASEC: It is even hard to imagine how fast this is (PRASEC = 3.3 light years and MEGAPRASEC = million x 3.3 light years).

In comparison, our farthest space probe from Earth, the Voyager 1, is travelling a million miles a day approximately, and after 40 years, the spacecraft has barely left our solar system. But everything is moving, nothing is static.

In fact, as you are reading this, the Earth revolves on its own axis; and orbits the sun, the sun moves through space at staggering 792,000 kilometers per hour around the galactic centre, and our universe moves at a mind-bending 2.1 million kilometers per hour.

In fact our universe is being pushed by MYSTERIOS forces in the universe (or out side the universe).

A new study recently published in Nature Astronomy describes previously UNKNOWN, supermassive region in our extragalactic neighbourhood largely empty of galaxies, which is exerting a REPELLING FORCE on our local group of galaxies.

Previously, experts thought that our galaxy was speeding through the universe because it is being pulled to a specific, dense part of universe due to that region's gravity. However, a new study has shown differently.

It turns out that the Milkyway is being sung across the universe by so called EXTRAGALACTIC VOID; in a process scientists call a kind of galactic tug of war, at a bending speed of 1.2 million miles per hour.

It wasn't until recently that scientists managed to figure out where we are located in space. Our cosmic neighbourhood. According to Astronomers, Laniakea Supercluster is the galaxy supercluster that is home to the Milkyway and 100,000 megasparses (520 million light years). It has approximate mass of 10^{27} solar masses, or a hundred thousand times that of our galaxy, which is almost the same as that of Horologium Supercluster. (Horologium Supercluster is also known as Horologium-Recticulum Supercluster, consisting of SCI 48 and SCI 49 is a massive supercluster, spanning about 550 million light years, it has a mass of 10^{17} solar masses, similar to that of Laniakea Supercluster that houses the Milkyway. It is estimated on coordinates right ascension 03^h and declination -50°02¹, and spans an angular area of 12° x12°). It consists of four subparts, which are known previously as separate superclusters.

But that's small if you zoomout further away. There are 10 billion galaxies in the observable universe! The number of stars in galaxies varies, but assuming an average of 100 billion stars per galaxy means that there are about 1,000,000,000,000,000,000 (that is billion trillion) stars in the observable universe.

Plate I

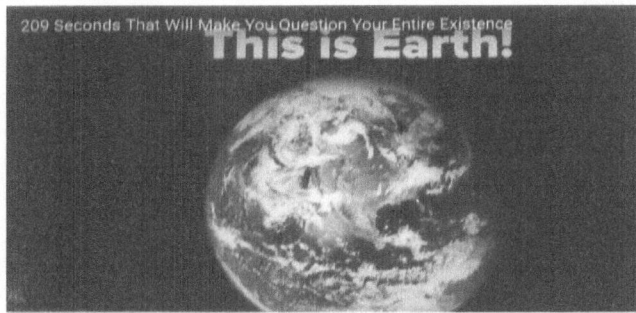

Plate II

Plate III

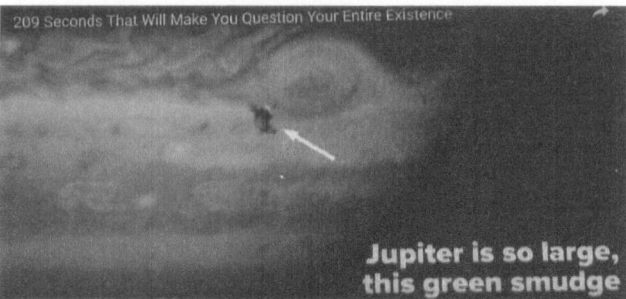

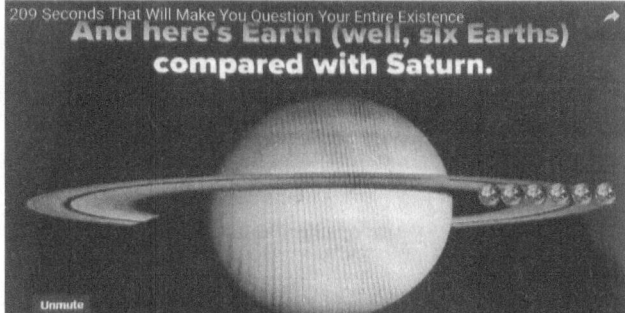

Plate IV

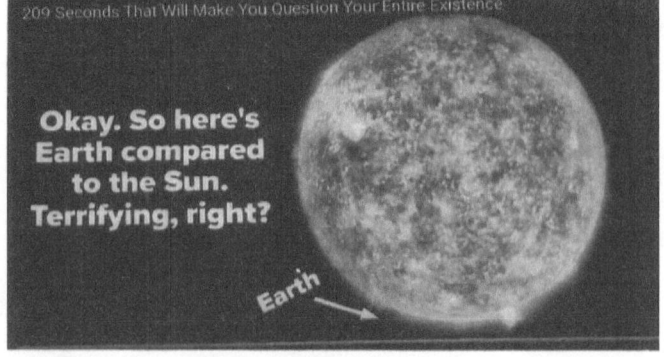

Plate IV

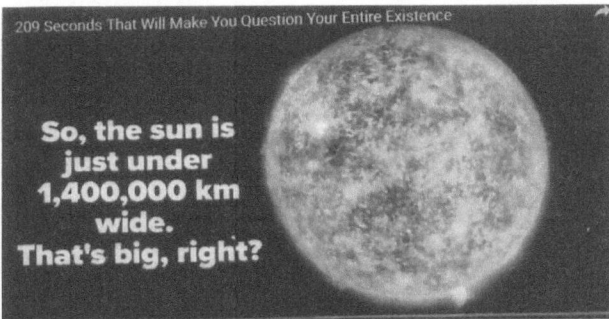

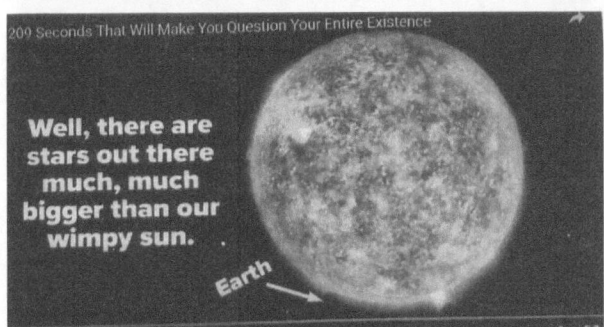

Plate V

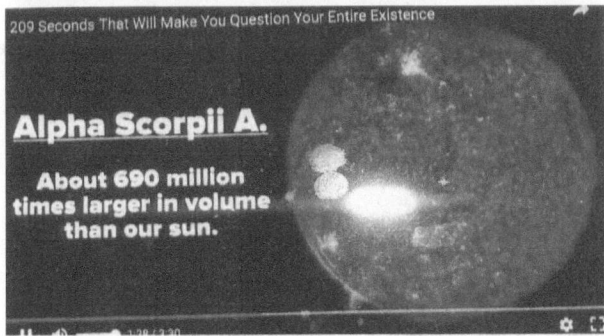

Plate V

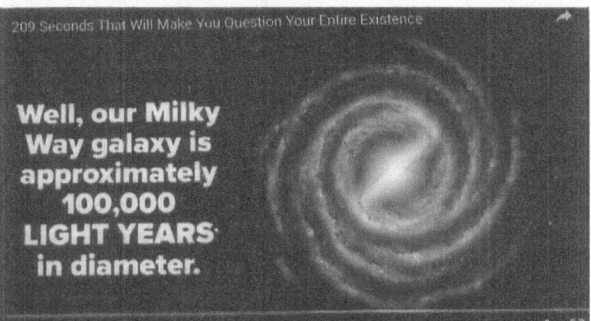

Plate VI

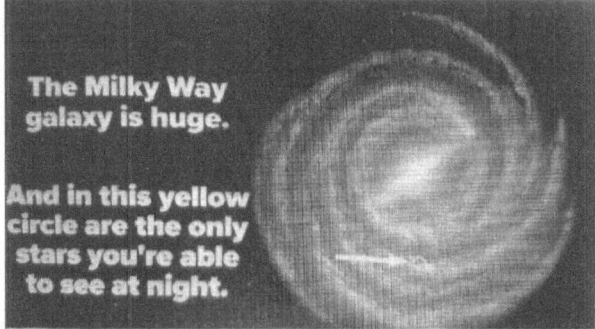

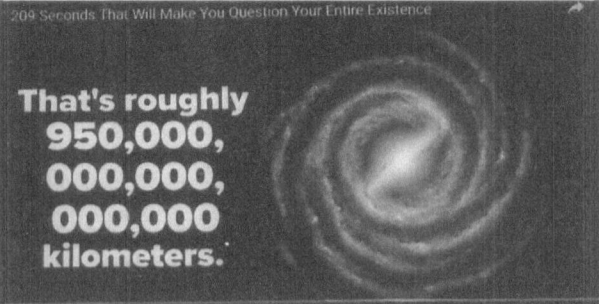

Plate VI

Plate VII

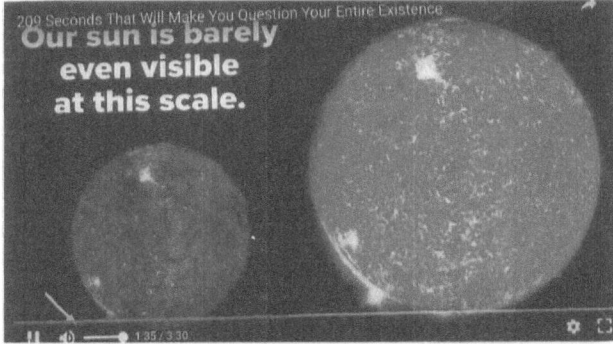

Plate VII

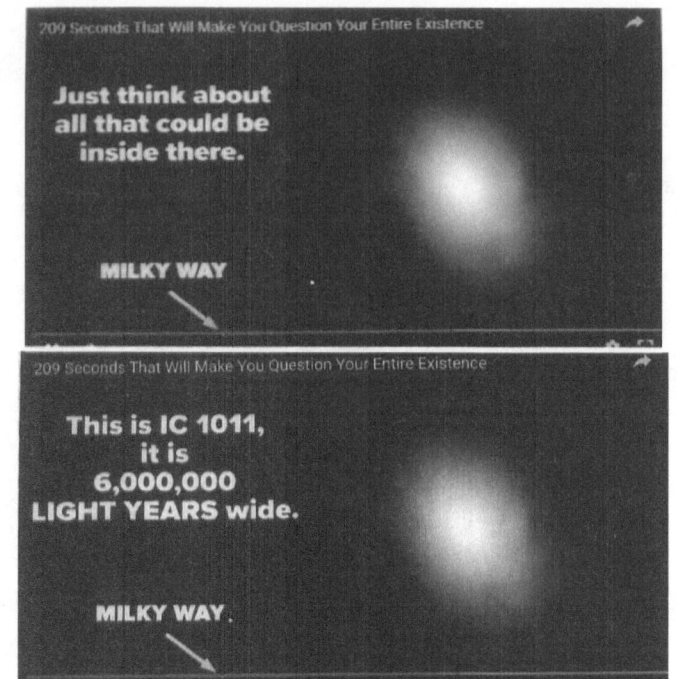

CHAPTER 15

THE BIBLICAL DOCTRINE OF CREATION

by Rev Dr. G. Wright Doyle

INTRODUCTION

The importance of this topic

Hardly any question could be of greater importance than that concerning the origin of the universe, including mankind. Virtually everything else depends upon how we answer this question. Our view of humanity – our origin, nature, purpose – determines how we see and treat ourselves and others. How we care for, or abuse, the environment flows from how we see the world. Naturally, all faith systems, including atheistic materialism, start with what we think about the origin, nature, and purpose of the world.

Within the Christian faith, likewise, virtually every other doctrine hangs upon the doctrine of creation. The Christian understanding of God, Christ, mankind, salvation, ethics, and the ultimate future, are all founded on the biblical teaching on creation. Who is God? "The Maker of heaven and earth," says the Apostles' Creed. Who is Christ? The one "by Whom are all things made," declares the Nicene Creed.

In particular, the Bible states that the sin of Adam, the first man, implicated all succeeding generations of men and women. In him, as our federal head, we all "died" spiritually, and will die physically.[23] On the other hand, all who believe in Christ will be "made alive" spiritually, and eventually raised to eternal life in a renewed earth.[24] If Adam was not an historical figure, and his death because of sin an historical event, then the death of Jesus Christ has no meaning and no effect on his followers.

The assault on the historicity of Genesis in the nineteenth century began what became a general attack on the reliability of the entire Bible. By the middle of the twentieth century, almost all "mainline" Protestant seminaries taught that the Bible is filled with errors of science and history, and therefore unreliable as a guide to faith and practice. That view has since invaded conservative Protestant seminaries also.

Other scholars have challenged this denial of the trustworthiness of the Bible, of course. Over the past several decades, a veritable flood of solid conservative scholarship has demonstrated, over and over again, that the Bible is historically accurate.[25] More recently, Christian scientists, along with non-Christians, some of them atheists, have begun to show how scientifically unsound the theory of evolution really is.

In this chapter, we shall first examine what the Bible says about the creation of the world, and then consider objections that have been raised against the biblical doctrine.

1. What the Bible says

Creation by God

The Bible says that God created the universe. Let us unpack that statement a bit. Who is this God and what is he like?

---

[23] Romans 5:12, 15-19

[24] Romans 5:15-21

[25] Carl Henry presented the most extensive defense of the reliability of the Bible, but thousands of other scholars have provided overwhelming evidence that the Scriptures teach true history. See Carl F. H. Henry, *God, Revelation and Authority*, Six Volumes, (Waco, TX: Word Books, 1976-1983)

God is eternal. "In the beginning, God created the heavens and the earth."[26] Thus begins the entire Bible. In other words, before the beginning of time and history, God existed.

God is a Spirit, not a field of energy or eternal matter.[27] Thus, he is invisible. No one has ever seen his essence.[28]

God is all-powerful in every way. He is the possessor and source of all energy, information, life, and everything else that can accomplish any kind of "work," broadly defined.[29]

God is unitary, but also plural, and thus personal. This is hinted at in the first sentence of the Bible, quoted above, in which the subject, "God," in Hebrew is plural, and the verb, "created," is singular. Later, God says, "Let us make man in our own image… So God made man in his own image, in the image of God he created him; male and female he created them."[30] Mankind, as both male and female, somehow reflects both the unity and plurality of the one God. The New Testament developed this idea much more clearly. There, we find references to "God"; the "Word," or "Son" of God, who is also called "Lord," meaning that he is equal to "God"; and also the Holy Spirit.[31] This, of course, is an unfathomable mystery, but it is taught in the Bible, and explains some things otherwise inexplicable, as we shall see.

God is rational. He has a mind, which expresses itself in speech. This "Word" of God, also identified as the Son of God who later became the God-man Jesus, served as the active agent of creation. "God said, 'Let there be light,' and there was light."[32] The other acts of creation are described in the same way: "By the word of the LORD the heavens were

---

[26] Genesis 1:1
[27] John 4:24; Revelation 1:8
[28] John 1:18
[29] Revelation 1:8
[30] Genesis 1:26-27
[31] Matthew 28:19; John 1:1-3; 2 Corinthians 13:14; and many other passages.
[32] Genesis 1:3

made... For he spoke, and it came to be."[33] "In the beginning was the Word, and the Word was with God, and the Word was God... All things were made through him."[34] This Word, or Logos, of God, is the center and source of all information, mind, rationality, and speech. And he is personal, not an impersonal force or principle like one of Plato's Ideas.[35]

God rules the universe now by his Word-Logos. Contrary to the "divine clockmaker" theory, the Bible states that God continues to govern all aspects of the created order through his Word. That means that we are not dealing with impersonal "laws" of nature, much less random processes. No, the constants of the world, as well as the constant variations and developments, including motion, growth, and decay, are directed by the divine Logos, who is the Son of God.[36]

God created the world out of nothing. Matter is not eternal or self-existent; neither are energy or information. These were all brought into being by Almighty God *de novo*, fresh, from no pre-existing state, at the "beginning" of time and history. In other words, they all "came to be," in distinction from the divine Word, who always "was."[37]

God created the world according to an orderly plan. What we see is not the result of random, undirected chance events, but the product of an intelligent Designer. The opening chapter of Genesis states this clearly. At first, the universe was "formless and empty." Then, step by step, God brought "form" by separating things from each other (light from darkness; day from night; the waters above the earth from the waters below; the sea from the dry land). Then he "filled" the world with vegetation; heavenly bodies; fish and birds; land animals of all sorts; and finally mankind.[38] The creation of light on the first day implies all the

---

[33] Psalm 33:6, 9
[34] John 1:1, 3
[35] See also Robert Gangke, ***Origins and Destiny: A Scientist Examines God's Handiwork***(W. Publishing Group, 1990).
[36] Hebrews 1:3
[37] John 1:1, 3; Romans 4:17
[38] Genesis 1:2-28

The Biblical Doctrine of Creation

energy spectrum, with visible light just a part. X-rays, gamma rays, and similar forces came into being with the creation of light.

God created living things according to "kinds."[39] These "kinds"(the Hebrews word is *baramin*) are not exactly the same as what we call "species," but they are self-reproducing kinds that cannot mix with other kinds to produce reproducing "offspring" (widely defined, to include plants and animals). Nor does one "kind" develop into another "kind." They are stable. [40]

God created men and women in his own image.[41] They possess some "divine" characteristics, such as mind, the ability to communicate, will, affections, and personality. Humans did not evolve from apes or any other "lower" form of animal life.

God created the world in six 24-hour days. The Genesis record leaves not room for doubt on this point. Each day is marked with a number, "the first…second…third" day, until the end of the sixth day. Then it says: "And there was evening and there was morning, the sixth day. Thus the heavens and the earth were finished, and all the host of them. And on the seventh day God finished his work that he had done, and he rested on the seventh day from all his work that he had done…in creation."[42]

God finished his work of creation. In other words, the initial project of creation is completed, not ongoing. What we observe now is the result of creation, and later the Fall, constantly developing and decaying under God's providential direction.

God created the world about 6,000 years ago. This number comes from the genealogies in Genesis 5:1-32 and 9:28-10:32.[43]

---

[39] Genesis 1:21, 24, 25, 26.
[40] Kurt Wise describes the nature of "kinds" at great length. See Kurt P. Wise, *Faith, Form and Time* (Nashville, TN: Broadman & Holman, 2002), 108-133.
[41] Genesis 1:26-27.
[42] Genesis 1:31-2:3
[43] Even allowing for some gaps in the genealogies, the world cannot be older than 10,000 years.

God created a good world. There were no defects; no discord; no disease, decay, or death.[44]

God created the universe to reflect his own beauty and goodness. The Bible often uses the word "glory" to express these two concepts of ethical and aesthetic goodness and excellence.[45]

Not long after the creation, Adam and Eve, the first man and the first woman, disobeyed God. As a result, they were condemned to spiritual "death," that is, alienation from God; alienation from each other; shame; eventual physical death; pain in childbirth for the woman; frustration at work for the man; enmity between humans and at least some animals (notably the serpent); and a general "curse" on the earth.[46] The disorder, disintegration, disease, and death we see around us – in other words, the outworking of the Second Law of Thermodynamics – derive from this initial catastrophic event.[47]

About 4,300 years ago, another catastrophe affected the entire globe – the universal Flood of Noah's day.[48] The "fountains of the great deep burst forth, and the windows of the heavens were opened."[49] The ensuing deluge covered the entire earth, including the highest mountains at the time, which may not have been as high as the highest peaks today. All humans except Noah, his wife, his three sons, and their wives, and the animals, insicts, and birds taken into the Ark, were exterminated. The

---

[44] Genesis 1:10, 12, 18, 21, 25, 31. Carl F. H. Henry explores the biblical doctrine of creation at great length in *God,Revelation and Authority,* Volume VI, Chapters 5-9.

[45] Psalm 19:1; Isaiah 43:7. The eighteenth – century American theologian Jonathan Edwards wrote eloquently and elegantly about this in many places, including, "The End for Which God Created the World." See Brandon J. Cozart and G. Wright Doyle, eds., Peter R. Doyle, *Jonathan Edwards on the New Birth in the Spirit* (Durham, NC: Torchflame Books, 2017), for an introduction to Edwards' thought, including this central theme of his.

[46] Genesis 3:1-19

[47] Romans 5:12-21;8:19-23

[48] Genesis 6:13-9:23. See also Job 12:15; 22:16; Psalms 29:10; 104:6-9; Isaiah 54:9; Matthew 24:37-39; Luke 17:26, 27; Hebrews 11:7; 1 Peter 3:20; 2 Peter 2:5; 3:6.

[49] Genesis 6:11

Ark itself was large enough to hold all the species of that time, and was virtually unsinkable because of its box-like shape.[50]

Many of the geologic features found today can be attributed to this worldwide catastrophe and the smaller catastrophic events that took place in the hundreds, and even thousands, of years following.[51]

That is the traditional view of the Genesis record, and it was universally believed by Christians and Jews until less than 200 years ago, when Darwinian evolutionism and Lyall's uniformitarian geology began to gain influence.

Now, however, this traditional view encounters massive resistance from many quarters. Let us look at some of the main objections to it, and see how these objections might be answered.

2. Objections to the traditional view of Genesis

A. Philosophical objections

Atheists say that God does not exist. He could not, therefore, have created the world. To this I want to reply: Really? How do you know God's doesn't exist? If you say, "Science has proven this," then please tell me which experiment demonstrates God's non-existence, and in what publication it was printed. If you say, "God can't be seen or detected with the senses," Iwouldask, "So, does this mean that whatever can't be measured or detected by the senses does not exist? How do you know that is true?" The non-existence of God is an unproven assumption

"God is an impersonal force, not a personal being." If that is the case, how do you explain personality in human beings? The mind? Communication? A sense of right and wrong? Love?

---

[50] The dimensions of the Ark show that it could hold the equivalent of 155 standard railroad boxcars.
[51] For substantiation of these statements, see especially Kurt Wise, *Faith, Form and Time*, 179-224.

In science, the law of causality states that every effect must have a cause. An intelligent Creator makes better sense as the cause of everything than any other hypothesis.

B. Scientific objections

"To start with, no reputable scientist believes in a six-day creation." Really? How do you account for the fact that at least 10,000 scientists with doctoral in the United States alone degrees have expressed belief in "young-earth creationism"?[52] Or the fifty scientists who explained why they hold to the traditional biblical view in the book, *In Six Days*?[53] Perhaps you are not aware of the growing number of scientists, including non-Christians, around the world who have written books that debunk Darwinian evolutionism.[54]

Other objections come from diferent fields of science:

Geology

"The earth is old, as the geological formations, dated by radioisotopes, prove." First, geological formations do not prove that the earth is old. For one thing, there is not one instance of a complete "geological column" on the earth. In addition, the old-earth theory cannot account for many

---

[52] You can listen to the story of how the former Dean of the Medical School at Tulane University abandoned Darwinism and became a Creationist at: https://www.youtube.com/watch?v=pS5j3XccmUM.

[53] John F. Ashton, Ph.D., editor, *In Six Days: Why Fifty Scientists* Choose to Believe in Creation. (Green Frost, AR: MasterBooks, 2001). If widely read, this one book would demolish the claim that the evidence supports evolutionary theory and that young-earth creationism is unscientific; the reverse is true. I shall cite only a few of the many findings in this collection of arguments from scientists from a variety of fields.

[54] These include Michael Behe, *Darwin's Black Box: The Challenge to Biochemical Evolution* (New York: The Free Press, 1996); David Berlinski, *The Deniable Darwin and Other Essays*, Edited by David Klinghoffer (The Discovery Institute, 2010);William A. Dembski, *Intelligent Design: The Bridge Between Science & Technology* (Downers Grove, IL: InterVarsity Press, 1999), especially 112-114; Michael Denton, *Evolution: Theory in Crisis* (London: Burnett Books, 1985); Don DeYoung, *Thousands...Not Billions: Questioning an Icon of Evolution; Questioning theAge of the Earth* (Green Forest, AR: MasterBooks, 2005); Jonathan Wells, *Icons of Evolution: Science or Myth? Why much of what we teach about evolution is wrong* (Washington, DC: Regnery Publishing, 2000). See also the volumes by Gangke and Wise, mentioned above.

anomalies, such as the presence of fish fossils on the peaks of high mountains. Young-earth creationism, however, can explain much more of what we see. KurtWise treats this matter at length.[55]

Secondly, dating by radioisotopes is highly unreliable. It depends on three assumptions: We know how much of the element was in the original "mother" rock; we know that no other forces or factors, such as the introduction of new elements, have altered the original composition of the mother rock; we know that the rate of nuclear decay has been constant. In fact, however, we cannot know how much of any element was present in the mother rock; we usually cannot guarantee that nothing has been added over the years, and there is strong evidence that the rate of nuclear decay has changed over the centuries.

Moreover, old-earth geology cannot explain why there are traces in Carbon 14 in ancient rocks – something impossible in today's world – or helium retention in zircon crystals. Nor can it explain why different dating systems assign vastly divergent dates to the same rock, or why demonstrably recent volcanic rocks have been given "old" dates. In fact, dating rocks and sedimentsis based on circular reasoning: The rocks are dated by their position relative to other rocks, and the age of these rocks is determined by evolutionary theory, the very thing the rocks are supposed to prove![56]

Uniformitarian geology that assumes billions of years suffered a heavy blow when the eruption of Mt. St. Helens in the state of Washington created formations that resemble those usually dated as very old.

"No steady-state river flow could possibly cover such a vast area" as the Great Basin of the western United States. "Neither would it produce the violently buried and mangled bodies found fossilized in many rocks of the region. The present-day erosion conditions applied uniformly

---

[55] See Wise, *Faith, Form and Time*, 181-220.
[56] For detailed discussions of these and other problems with dating by radioisotopes, see DeYong, *Thousands, Not Billions*. See also Paul Geim in Ashton, ed., *In Six Days*, 59-60.

in the past could not account for the unusual formations of the Grand Canyon, mesas, badlands, and other canyons."[57]

Biology

According to Darwinian theory, life evolved from the simple to the complex, by random mutation guided by natural selection, as fossils and examples of helpful mutation demonstrate.

First we must make the crucial distinction between minor changes within species and origin of new species. The former, often called "micro-evolution," has long been known and is never denied. In fact, that is what the small and temporary changes in the beaks of Darwin's finches on the Galapagos Islands illustrate.

The latter, called "macro-evolution," has never been observed in the laboratory, despite countless experiments to engineer a change from one species to another. Nor is it found in the fossil record, contrary to some famous and often-repeated "examples."[58] Even the horse series "is not as perfect as is commonly assumed…" and it does "not involve a great deal of change."[59]

Furthermore, multiple problems plague the fossil record as any proof of Darwinism or an old earth. I shall offer only a few examples:

- The so-called "Cambrian explosion" of millions of species of living creatures is supposed to have occurred hundreds of millions of years ago, but these living organisms appear suddenly, fully developed, with no previous forms existing. Furthermore, "modern" creatures are there, along with some that are now

---

[57] Jeremy Walter in Ashton, ed., *In Six Days*, 12-13.
[58] Jonathan Wells examines and explodes these so-called "examples" in *Icons of Evolution*. Wells summarizes his book, in which he shows that the most commonly used "examples" of evolution are in fact either frauds or failures, in a brief video: https://www.youtube.com/watch?v=xZn7tTdCm6U.
[59] Denton, *Evolution*, 184, 185. Denton devotes an entire chapter to the fossil record, in which he demonstrates that transitional forms between major divisions of living creatures are few and far between; the gaps remain.

extinct. Most importantly, complexity is already present; there is no development from the "simple" to the "complex."

- Indeed, as Michael Behe has shown, there is an "irreducible complexity" in all forms of life. That is, each cell has an incredible collection of parts, all of which must be present and functioning for the cell to exist and do its job in the organism.[60] Darwin lacked the instruments that would have enabled him to see this complexity, but it is now observed daily in laboratories all over the world, and cannot be denied. Even Darwin himself could not explain the complexity of the human eye, which he admitted must be complete at all times for it to function at anytime.[61] The amazing complexity of what is now called Biodiversity also demands that all parts work together at once.[62]

- Soft tissue has been found in fossils of dinosaur bones, which are supposed to be millions of years old. When scientists see this soft tissue, including blood vessels, they are incredulous.[63] The conclusion: Dinosaurs cannot be more than 10,000 years old. Carbon 14 has likewise been discovered in oil deposits and coal deposits. The existence of Carbon 14 in such supposedly "old" mineral deposits is impossible, unless the coal and oil formations are not as old as evolutionary geology maintains.

- The transitional forms which Darwin expected would be found have never shown up. They are still "missing links." In other words, the fossil record contains no instances of transitions from one "kind" to another. The well-known examples of horse and human bones offer no proof, either.[64] In the words of Michael

---

[60] Behe, *Darwin's Black Box*, 39-122. Other discussions of the fundamental complexity of the cell and "lower" forms of life may be found in Wise, *Faith, Form and Time*, 99-107; John K.G. Kramer, in Ashton, ed., *In Six Days*, 48-49.
[61] For more on complexity, see Jerry Bergman in Ashton, ed., *In Six Days*, 24-34.
[62] See Henry Zuill, in Ashton, ed., *In Six Days*, 63-74.
[63] You can witness their reactions at https://www.youtube.com/watch?v=WdqYPjA9VxA ;
[64] See Wells, *Icons*, 195-228.

Denton, more than a century of examination of fossils has "never yielded...any of Darwin's myriads of transitional forms...the infinitude of connecting links [posited by Darwin] has still not been discovered and the fossil record is about as discontinuous as it was when Darwin was writing... The intermediates have remained as elusive as ever and their absence remains...one of the most striking characteristics of the fossil record."[65]

Other, more fundamental, problems call the theoretical basis of evolutionary theory into question. For example:

Darwinian evolution holds that new species have resulted from innumerable changes in organisms over billions of years to produce new forms that are more adapted to the environment. The proposed mechanism for these changes is natural selection, working through random mutations. But

- Almost all mutations are destructive. Even beneficial ones make only minor transformations within a species. In other words, mutations *cannot* produce improved species; they cannot produce new species, in the sense of biblical "kinds." These facts are known by all biologists.

- The odds of random beneficial mutations – even if they could take place – of producing improved life forms have been computed by eminent scientists and mathematicians. Mathematician David Berlinsky, for example, demonstrates that these odds are impossibly great for even the smallest changes in the tiniest units of life.[66] Information scientist William A. Dembski does the same from a statistical viewpoint.[67] Astronomer Fred Hoyle Hoyle used the illustration comparing the random emergence of even the simplest cell to the likelihood that "a tornado sweeping through

---

[65] Denton, *Evolution*, 162.
[66] Berlinsky, *Deniable Darwin*.
[67] Dembski, *Intelligent Design*, 127-139, and elsewhere. See also Jerry R. Bergman in Ashton, ed., *In Six Days*, 32-40.

a junk-yard might assemble a Boeing 747 from the materials therein."[68] Hoyle also likened the chance of obtaining even a single functioning protein by chance combination of amino acids to a solar system full of blind men solving the Rubik's Cube simultaneously.[69] Evolutionary molecular biologist Michael Denton, though an atheist, demonstrates the impossibility of random mutations as the mechanism of beneficial mutations leading to the creation of even the simplest forms of life in his chapter, "Beyond the Reach of Chance."[70] More recently, biochemist Michael Behe has described the impossibility of random helpful changes.[71]

In sum: Random mutations cannot produce new species (defined as biblical "kinds"), nor can they result in life forms that are more adaptive to their environment. These irrefutable facts of science strike at the foundation of evolutionary theory.

Darwin himself stated candidly, "If it could be demonstrated that any complex organ existed which could not possibly have been formed by numerous, successive, slight modifications, my theory would absolutely break down."[72] Well, this has been abundantly demonstrated. The foundation is shaking.

A number of facts in chemistry deny the theory of evolutionary chemistry.[73]

Fundamentally, the theory of evolution comes to shipwreck upon the immovable shoal of the Second Law of Thermodynamics. This proven law of science holds that throughout the universe, there is a universal process of a "decrease of energy available to effect further processes.

---

[68] Hoyle, *Nature* **294**(5837):105, November 12, 1981.

[69] Hoyle, "The universe: past and present reflections," *Engineering and Science*, p. 12, November 1981.

[70] Denton, *Evolution*, 3 08-325. See also Behe, *Darwin's Black Box*, 29;

[71] Behe, *Darwin's Black Box*, 144-161. He concludes, "Molecular evolution is not based on scientific authority." 185.

[72] Charles Darwin, *Origin of Species*, 6th edition (New York: New York University Press, 1988, 154; original edition 1872) cited in Behe, *Darwin's Black Box*, 39.

[73] See Jonathan D. Sarfati, in Ashton, ed., *In Six Days*, 81-85.

Alternatively stated, real processes result in a net increase in the 'entropy' of the universe."[74] In other words, things fall apart, decay, disintegrate, and eventually die. I put it this way, "Cookies crumble." More importantly for biological evolutionism, as Gangke shows, is that the Revised Second Law defines this degenerative process more precisely to state that there is a decreasing amount of useful information available. Evolution requires new information, producing new life forms that are better than their forebears. The Second Law makes this impossible.[75]

Physical Anthropology and Palaeontology

"Mankind descended from apes, as the fossils show." The problem is that the fossil record does not demonstrate this assertion.[76] Some supposed "links" in the evolution of modern man from earlier ancestors, like the Java Man and the Piltdown Man, have turned out to be fakes.[77] Others are now properly classified as *Homo erectus* or even *Homo sapiens*.[78] Even Peking Man, much-beloved by the Chinese, is surrounded with mystery.[79] Others, like Leaky's "Lucy," cannot be shown to demonstrate any relationship to modern humans.[80] In any case, the problems with dating that plague the use of radioisotopes to assign different ages to rocks mentioned above render any claim for human-like figures to be more than 10,000 years old questionable.

---

[74] Jeremy L. Walter, in Ashton, ed., *Six Days*, 15.

[75] Gangke, *Origins and Destiny*.

[76] See especially *Bones of Contention: A Creationist Assessment of Human Fossils, Revised Edition* (Grand Rapids, MI: Baker Books, 2004). In addition, see https://answersingenesis.org/human-evolution/ape-man/is-there-really-evidence-that-man-descended-from-the-apes/; https://answersingenesis.org/human-evolution/hominids/cranial-vault-thickness-of-homo-erectus-and-modern-man/.

[77] On Java Man, see https://answersingenesis.org/human-evolution/hominids/who-was-java-man/. The well-known falsity of Piltdown Man is explained at https://en.wikipedia.org/wiki/Piltdown_Man.

[78] For a brief discussion, see Wise, *Faith, Form and Time*, 232-238.

[79] For one thing, the bones have disappeared. http://io9.gizmodo.com/5980372/the-bizarre-disappearance-of-the-peking-man-fossil. In any case, Peking Man is now recognized as *Homo erectus*.

[80] For a lengthy article on various attempts to infer the evolution of modern humans from fossil remains like "Lucy,", see http://www.christiananswers.net/q-aig/aig-c029.html

Astronomy

"The universe is vast and old, as 'light years' prove." Distant starlight, proponents of an old universe claim, confirm that at least the stars are billions of years old.

This objection appears to be very strong, but it, too, faces challenges.

For one thing, the "Big Bang" theory, which currently stands as the main alternative to creation by an Intelligent Designer, has to explain several improbable developments:

- From disorder to order. In everyday life, explosions create chaos, not order. Fred Hoyle's analogy, mentioned above, applies here also.
- From non-life to life. This progression has never been observed, even in controlled laboratory conditions.[81] Indeed, the odds against the spontaneous, undirected production of life from non-life are impossible.[82]
- From no information to information.

Observationally, the Big Bang theory, which calls for different temperatures at different points in the universe, must explain why temperatures are mostly the same throughout the universe. This is the so-called "Horizon Problem" of the theory, and it has never been solved.[83]

The Big Bang theory assumes that the universe is at least four billion years old. The moon, however, is known to have receded from the earth at the rate of about an inch and a half every year. If the solar system is only 6,000 years old, that is no problem, but at the current rate of recession, the moon about have been touching the earth as "recently" as 1. 5 billion years ago, which is fatal to the Big Bang Theory.

---

[81] The false inferences drawn from famous Miller-Urey experiment have been refuted in many places, including Wells, *Icons*, 9-27.
[82] See Jerry R. Bergman, in Ashton, ed., *In Six Days*, 40-43.
[83] Jason Lisle, *Taking Back Astronomy* (Green Forest, AR: MasterBooks, 2006), 48-50. This and the examples following come from Lisle. I am not in a position to evaluate their cogency, but cite them as several among many objections to the Big Bang theory.

Nor can the Big Bang theory explain the virtual absence of antimatter (matter with a negative charge) in the universe. According to the theory, there should be an equal amount of antimatter and matter.

Most basically, the Big Bang theory cannot answer the question of origins: where did the original elements come from? What ignited the explosion? What guided the ensuing processes leading to the incredibly complex, yet intricately ordered, universe?

Apart from the Big Bang theory, old-age theories of the earth face other problems: The earth's magnetic field is weakening by about 5% every century. 6,000 years ago, it would have stronger than it is today, but still within the range of viability for life on earth. On the other hand, if the earth were as old as many scientists claim, the magnetic field would have been so strong that life on earth would not have been possible.

Another problem is the existence of spiral galaxies, which slowly rotate, "becoming more and more twisted up as the spiral becomes tighter. After a few hundred million years, the galaxy would be wound so tightly that the spiral structure would no longer be recognizable."[84] But we do see such galaxies today, more evidence of a recent creation.

Comets develop a "tail" as they lose material. Astronomers generally estimate that comets cannot last more than 100,000 years at most, but we still see many comets in the solar system, another indication of a recent creation.[85]

Secular models of solar development call for the smaller, rocky planets to orbit more closely to their star, while the larger, gas planets orbit farther away, as in our Solar System. The difficulty is that most of the 150 planets orbiting other stars that have been discovered fail to fit this model: the giant gas ones orbit very close to their star.

---

[84] Lisle, *Astronomy*, 66.
[85] The supposition of the existence of an "Oort cloud," composed of a vast reservoir of icy masses, some of which falls into the solar system to form new comments, is without scientific basis. It is only a theory invented to solve the problem of the existence of comets.

Star formation cannot be explained by the models which pre-suppose an old universe, because "gas pressure, angular momentum, and magnetic fields all work against the possibility of a condensing star."[86]

Even distant starlight does not prove an old age for the universe, for it rests upon the assumption, disproven by Einstein, that the speed of life remains constant.[87] Russell Humphreys has developed an alternate model based on Einstein's General Theory of Relativity. [88]

The uniqueness of the earth, and its incredible suitability not just for life, but for human life, argue against a random, undesigned formation of the universe, including a Big Bang. Michael Denton has documented some the amazing factors that make Planet Earth uniquely suited for human life.[89]

Fundamental Flaws in Secular Scientism

There are at least two fundamental flaws in any argument from science purporting to prove that the world was not created by God, but by purely materialistic and random processes:

First, science deals with observable phenomena. Science can only "prove" what is repeatedly observed, whether in the world or in the laboratory. But the origin of the universe happened only once. Furthermore, no one was there to observe it. This question, then, is not one falling within the purview of science.

---

[86] Lisle, *Astronomy*, 85.
[87] For a brief treatment of the question of distant starlight, see Chaffy & Lisle, *Old-Earth Creationism*, 140-143.
[88] See Russell Humphreys, *Starlight & Time* (Green Forest, AR: MasterBooks, 1996). Humphreysresponds to criticisms of the model by Hugh Ross at http://creation.com/dr-humphreys-responds-to-criticism-of-his-book-starlight-and-time. An updated interview with Humphreys can be found at http://evidencepress.com/dr-russell-humphreys-on-starlight-and-time/. More recent repsonses to criticisms can be obtained at https://www.trueorigin.org/ca_rh_03.php.
[89] Michael Denton, *Nature's Destiny: How the Laws of Biology Reveal Purpose in the Universe* (New York: Free Press, 1999}, Part I.

The study of the past is called History. Any particular history is credible or not, depending on whether the records are reliable and whether the original witnesses to events are trustworthy and give a convincing testimony to what happened. The Bible claims to be history; science cannot in any way judge whether the Bible's account of the origin of this world is correct.

Second, most of today's secular science is based on unproven assumptions that. The first assumption excludes, in principle, the possibility of a God, much less a creator God. Evolution and its cousins assume that, since God does not exist, he must not have played any part in the creation of the world. This is an unproven assumption, however. By the nature of the case, it cannot be tested, at least not theoretically; it just *is*. If you believe it, you interpret all evidence in its light.

The second assumption is called "uniformitarianism," that is, that everything has happened at the same rate from the beginning of time until now. There is no evidence for this assumption, however, and much evidence (such as catastrophic geological events) against it.

What if, however, much of the evidence contradicts your assumption? And what if another assumption – that God created the world, for example – offers a better explanation of the observed phenomena? You are then faced with a choice. This is a matter of philosophy based on faith, not science, as legal scholar Philip Johnson, along with many others, has demonstrated.[90]

At the beginning, the rejection of biblical creationism issued from the desire of some leading British intellectuals, like Thomas Huxley, to remove God from the arena of science and to prove that he had no part in creating the world. They began, that is, with an agenda and an assumption. The majority of evolutionary scientists today labor with the same anti-god agenda and under the same assumption.[91]

---

[90] See Philip E. Johnson, *Darwin on Trial* (Downers Grove, IL: IVP Books, 2020).

[91] For the militant atheism of the "scientific establishment" in the United States, see Jonathan D. Sarfati, in Ashton, ed., *In Six Days*, 76-77.

Indeed, as I said above, the fundamental issue is one of worldview and assumptions, or philosophy, broadly defined, not one of science. Science can have nothing to say about the origin of the universe or of life. In fact, the philosophy of science has gone through many developments, and is in constant flux. Even more transitory are the so-called "findings" of scientific research. As Carl Henry never tired of pointing out, scientists constantly revise their theories, hypotheses, interpretations, and conclusions. That, of course, is as it should be, if science is to remain true to its nature as ongoing inquiry based on constant observation and inference. Evolutionary theory is no exception. In a *tour de force* of apologetics, Carl Henry demonstrated the embarrassing confusion and contradictions among proponents of evolutionary theory.[92]

This entire debate largely rests upon alternative assumptions, and is really a matter of philosophy and history, not science. Though I am not a professionally trained philosopher and am certainly not a scientist, I am not totally unexposed to philosophy, and I think I can spot an assumption when I see it.[93] Illegitimate logical jumps and circular reasoning are also easy to detect. Even a layman can see that arguments for evolution are filled with unproven assumptions and faulty logic.

C. Biblical objections

Under the onslaught of what appeared at the time to be the "assured results" of science, beginning in the late 19th century, some biblical scholars and theologians tried to find ways to re-interpret the Genesis record to accommodate what they thought was clear evidence from science that the traditional interpretation could not be right. In doing so, they departed from more than 1900 years of Christian

---

[92] See Carl Henry, *God, Revelation and Authority,* Volume VI, Chapter 8, "The Crisis of Evolutionary Theory," 156-196.

[93] One of my doctoral exams was on Hellenistic philosophy, and my work on historical theology and on Carl Henry, who held a Ph.D. in Philosophy, has brought me into contact with various philosophical developments in the West.

interpretation, and an even longer history of Jewish understanding of their own Scriptures.[94]

Let me say at the outset that most of these scholars are competent students of the Bible or of science, and hold to orthodox Christian beliefs. What follows does not in any way mean to impugn their integrity or intelligence, but to question their assumptions, methods and conclusions.

Here are a few of their attempts:

- "A long 'gap' of perhaps millions or billions of years can be inferred from the tenses of the verbs in Genesis 1:1-2. Thus, even if we assume the historicity of some sort of 'six-day' creative work, that took place after ages of time." First proposed in 1814, this theory has gained wide popularity among conservative Christians.

  The "Gap Theory" fails to account for the unambiguous narrative flow that beings with, "In the beginning God created the heavens and the earth," in Genesis 1:1 to "And God saw everything that he had made, and behold, it was very good. And there was evening and there was morning, the sixth day. Thus the heavens and the earth were finished, and all the host of them. And on the seventh day God finished his work that he had done, and he rested on the seventh day from all his work that he had done" (Genesis 1:31-2:3).

What could be clearer? God "created the heavens and the earth" (1:1), including "everything that he had made," in six days (1:31). He had "finished" making "the heavens and the earth, and all the host of them" (2:10) – that would include the stars and planets – and then he rested. This statement is made three times in two verses, a repetition that must mean something! All that God made, he made in six days. Everything.

---

[94] Contrary to Hugh Ross's claims, almost all Early Church Fathers stated that the world was young and that the days of Genesis 1:1-2:3 were twenty-four days. See Mark Van Bebber and Paul S. Taylor, *Creation and Time: a report on the Progressive Creationist book by Hugh Ross* (Mesa, AZ: Eden Productions, 1994), 93-102. The same goes for the Reformers, who were all expert biblical scholars. See John L. Thompson, ed., *Reformation Commentary on Scripture. Old Testament. I. Genesis 1-11* (Downers Grove, IL: InterVarsity Press, 2012), 1-70.

There was no "gap," much less an interval of billions of years, in the process. Any other reading of this passage defies all accepted norms of literary analysis and biblical exegesis.[95]

- "The word 'day' can mean 'age,' so the 'days' of Genesis 1:1-2:3 can be referring to ages of indeterminate length that would allow for the long processes assumed by Darwinian evolution, uniformitarian geology, and uniformitarian astronomy."

Now it is true that "day" (*Yom*, in Hebrew) can sometimes mean "period of time of indeterminate length," as in Genesis 2:4. On the other hand, "day" always refers to a twenty-four hour period of time when

- It is combined with an ordinal number, like, "first," "second," "third"
- It is associated with the word "morning," as in Genesis 1:5, 8, and throughout the chapter
- It is associated with the word "evening" as it is in Genesis 1
- When evening and morning occur together, as in Genesis 1
- When it is contrasted with "night," as in Genesis 1:5

There are no exceptions to these rules in the Hebrew Bible. That is to say, the word "day" in Genesis 1:1-2:3 can have only one meaning: an ordinary, twenty-four hour day.

- "The sun and moon weren't made until the fourth day. Thus no ordinary 'day' could have existed before then." This objection, however, overlooks the fact that a "day" is measured by the rotation of the earth on its axis, not upon the presence or absence of the sun.
- "The Bible is not meant to be a scientific textbook. We should not try to press modern questions or theories of science into an ancient narrative with primarily theological and ethical purposes." True, the Bible is not a scientific textbook. That misses the point,

---

[95] The grammatical structure of Genesis 1:1 and following also fails to support the theory, since it follows the so-called "*waw* consecutive" pattern used in continuous narrative.

however. As we said above, much of Scripture is history. Genesis 1:1-2:3, on any straightforward reading of it, purports to narrate the creation of the world by God during six consecutive days.

- "The creation account is 'poetry' with a 'framework' structure, with Days 1, 2, and 3 paralleled by Days 4, 5, and 6. Clearly, this parallel structure indicates that the writer was aiming to make a general claim, that the world was created by God, and is not eternal or the product of many of the 'gods' worshiped in ancient times. He was not trying to relate the precise order in which God made the world." To which we answer:
- The entire book of Genesis – indeed, almost every book in the Bible – is a beautiful literary work of amazing intricacy.[96]
- The parallels between the first three and the last three days of creation fit the author's description of how God first "formed" and then "filled" a world that was "formless" and "empty." This is straight historical narrative, not poetic parallelism.
- A close analysis of the finite verbs used in Genesis 1:1 to 2:3 demonstrates without the possibility of refutation that this passage belongs to the genre of history, not poetry. Simply put, the verbs are almost all in the preterite form, which is used in Old Testament historical sections, rather than the use of imperfect and perfect forms characteristic of poetry.[97]

In other words, the narrative of creation events in Genesis 1:1-2:3 is just that, a narrative, not a poem. This is an incontrovertible fact of literary analysis.[98]

---

[96] Umberto Cassuto demonstrated this brilliantly in A Commentary on the Book of Genesis, Part 1 (Magnes Press, 1978).

[97] For the rigorous statistical analysis done to demonstrate this fact, see the research findings of Steven Boyd in DeYong, *Thousands,...Not Billions*, 157-170.

[98] For an extended discussion of the hermeneutics and proper exegesis of Genesis 1:1-2:3, see Tim Chaffey 7 Jason Lisle, *Old-Earth Creationism on Trial: The Verdict is In* (Green Forest, AR: MasterBooks, 2008) 31-56.

May I add a personal note here? In 1959, at the age of fifteen, I began to read ancient texts in Latin. I continued with the close study of Latin texts, both poetry and prose, during college as a Latin major at the University of North Carolina, adding Greek in 1963. (I also spent a year studying French literature.) While in seminary, in addition to courses in New Testament Greek exegesis, I began learning Hebrew. In my second year of Hebrew studies, we examined both poetic and historical texts carefully. Later, I returned to UNC-Chapel Hill for four years of graduate studies in Classics, culminating in a doctoral dissertation on Augustine's rhetorical theory and how he applied it to his preaching. By that time, I had devoted eleven years to the close analysis of (mostly) ancient texts, both poetry and prose, under the direction of outstanding scholars. (This does not include the courses in English literature I took in high school and college.) I later taught New Testament Greek, including Greek exegesis, at a Chinese seminary in Taiwan for seven years.

Since about 2004, I have added history to my area of academic interest. As editor of the online *Biographical Dictionary of Chinese Christianity*,[99] co-editor of the series, *Studies in Chinese Christianity*,[100] and principal writer for the Global China Center website,[101] I have read dozens of volumes of serious history and have edited or authored almost two thousand pages of history of various kinds.[102] All of this has given me some idea of what constitutes historical narrative.

I hope you will forgive me for saying that I believe I know the difference between historical narrative and poetry. Genesis 1:1-2:3 is definitely history, not poetry. There are poetic descriptions of God's creation work in the Bible (like Psalm 104), as well as historical psalms (like Psalm 105). These are clearly poetry, however, and easily

---

[99] www.bdcconline.net

[100] http://wipfandstock.com/catalog/series/view/id/49/

[101] www.globalchinacenter.org.

[102] These include a 400-age history of Christianity in America (*Christianity in America: Triumph and Tragedy.* Eugene, OR: Wipt & Stock, 2013); reviews of books about the history of China and of Christianity in China; and many biographies of Chinese Christians and foreign missionaries.

distinguishable from the historical narratives of Genesis or the other historical portions of the Old Testament.

Many more similar objections on purportedly biblical grounds can be found in the writings of Hugh Ross, who is not a biblical scholar, as well as in books by respectable biblical scholars. Most of these are examined and refuted by Van Bebber and Taylor in the book cited in Note 66, above.

It seems to me (and to others) that the scholars who try to accommodate the Bible to modern science, even to the extent of forcing the text to say what it cannot mean, start from an incorrect assumption. They assume that the "findings" of modern science possess equal authority with the biblical text. Sometimes they talk of "two books" of God's revelation, one being the Bible and the other of nature, as if these two sources of revelation were on a par with each other.

The Bible, however, does not speak of two "books," but one – itself. Even Psalm 19, whence this theory derives, contrasts the written word of the Scriptures with the unspoken "word" of general revelation. For Protestants, there can be only one ultimate source of authority for faith, the Scriptures (*sola Scriptura*).

Secondly, the so-called "findings" or "assured results" of modern science, like those of negative critical biblical scholarship, continue to shift and change under constant revision. How can such an unsteady foundation support any doctrine, much less one as fundamental as the doctrine of creation? "Science," like many other disciplines, goes through phases and fads. Thomas Kuhn has shown that most scientists do their work based on a generally accepted paradigm, which "everybody" accepts as valid. Until, that is, some brave soul (like Galileo or Einstein) comes along and challenges the entire framework of the reigning orthodoxy. After a while, the old model seems hopelessly wrong, and people wonder how anyone could have believed in it.[103]

---

[103] See Thomas S. Kuhn, *The Structure of Scientific Revolutions*, 3rd Edition (Chicago: University of Chicago Press, 1996).

Many do not hesitate to use the word "myth" to describe evolutionary dogma today. Michael Denton, who does not accept the biblical model, boldly states, "Ultimately the Darwinian theory of evolution is no more nor less than the great cosmogenic myth of the twentieth century."[104] The time is coming when this modern myth will share the dustbin with the fables of ancient Greek, Roman, and Nordic gods. Darwin will join Marx and Freud among the great "pioneers" who have been massively discredited by the hard facts of reality.

A large and growing body of scientists today believe, and with good reason, that Darwinian evolution and the uniformitarian geology and astronomy that both flow from and support it, will soon collapse under the weight of its internal contradictions, massive contrary evidence, and the superior explanatory power of another paradigm. The only candidate for that "other" viewpoint is creation by an almighty and eternal God.

Therein lies the real obstacle to the acceptance of the biblical account. It's not the evidence for evolution, which they know is scanty, or the cogency of its theory, which is riddled with fatal scientific and logical flaws, but the knowledge that if they abandon Darwin and his descendants, they will have to face God and his truth.

So, why should sincere Christians, who trust in Christ and do not fear God's judgment, align themselves with atheists whose presuppositions lead to construct gossamer theories that will soon disappear?

Sadder still is the way that "mainstream" Christian publishers and academics relegate creationism, especially six-day creationism, to the margins of discourse.[105] Some churches will not even allow Sunday school classes in which both evolution and creation in six days are compared as different worldviews.

---

[104] Denton, *Evolution*, 358. See also Richard Milton, *Shattering the Myths of Evolution* (Park Street Press, 2000).

[105] That is why a number of books referenced in the notes in this chapter do not come from "mainstream" evangelical publishing houses; they are effectively banned, even though written by competent scientists.

CONCLUSION

The biblical record is a credible, rational position that explains what we see and experience, such as

The existence of matter, energy, information.

Personality, based on the creation of men and women in the image of a tri-personal God.

Complexity with order, including the laws of causality, resulting from creation by an intelligent Designer with a definite plan in mind.

Life, flowing from the God who has "life in himself,"[106] and from the agency of the personal Logos of God, "in whom was life,"[107] and who is himself "life."[108]

Beauty, which is the reflection of an infinitely beautiful and glorious Maker.

Rationality and the laws of logic, rooted in the Logos of God.

Moral consciousness, reflecting the goodness, holiness, and justice of the Creator.

Many phenomena on earth that cannot otherwise be explained, including the special position and situation of Planet Earth.[109] Others are "the population growth, the helium content in this world, the missing neutrinos from the sun, the oscillation period of the sun, the decline of the earth's magnetic field, the limited number of supernovas, radioactive halos, the mitochondrial DNA pointing to one mother, and the increase in genetic diseases.[110]

Some important observable astronomical phenomena, such as the magnetic fields of the planets in our solar system. For example, Dr. Russell Humphreys has developed a model, based on a 6,000-year-old creation,

---

[106] John 5:26.
[107] John 1:4.
[108] John 11:25; 14:6.
[109] Kurt Wise lists dozens of these in *Faith, Form, and Time*.
[110] John K.G. Kramer, in Ashton, ed., *In Six Days*, 53-54.

that predicted the strengths of the magnetic fields of Uranus and Neptune before they were measured by the Voyager spacecraft.[111]

Likewise, a multitude of phenomena in the world of living beings cannot be explained apart from intelligent design. These include the sonar system of the dolphin; insect flight; the amount of information in the simplest cell; the beauty and usefulness of the DNA code; "complex rotary motors in living organisms; the complex eyes of 'supposedly 'primitive' invertebrates"; lobster eyes; the sense of smell in humans.[112]

Both the dignity and depravity of mankind, who, though made in God's image, rebelled against his Maker and plunged the entire world into a descent into disorder, disintegration, and death. Based on the Genesis record, we understand why people are capable of both nobility and genius, on the one hand, and of barbaric behavior on the other. We also understand why Christians believe that all people need to be saved from our sins.

Another great miracle, made credible by the original miracle of creation, occurred when Jesus was raised from the death three days after he died upon the Cross. The Bible teaches that if we repent of our sins and trust in him, God will forgive us, restore us to fellowship with himself in this age, and in the age to come give us eternal life in a "new heaven and a new earth, in which righteousness dwells."[113]

I have touched upon only a few of many possible topics in this chapter. Thousands of learned tomes written by eminent experts in science, philosophy, mathematics, engineering, law, and the Scriptures have examined the cases for Darwinian theory and variations upon it, and the biblical account as traditionally understood. I hope that this sampling will provoke more investigation by open-minded seekers of the truth.

G. Wright Doyle

---

[111] D.R. Humphreys, "The Creation of Planetary Magnetic Fields," *Creation Research Society Quarterly* 21 (3) (December 1984), cited in Lisle, *Astronomy*, 64.

[112] See Jonathan D. Sarfati, in Ashton, ed., *In Six Days*, 78-81.

CHAPTER 16
CREATION AS PER ISLAM-HOLY QURAN

The Creation of the Universe
ISLAM
By: A. Muhammad

One of the oldest questions to be asked about the universe was "How did it begin?" This question always puzzled mankind. The answer to such a question depended on the faith and civilisation current to the scientist.

In the early days of the Greek philosophers, the universe was the making of gods and goddesses, but how they actually performed the act of creation was not to be inquired into, for such matters were considered divine and thus, laid outside man's comprehension. The Greek plan incorporated a marvellous scientific picture of the universe, for whilst they described the motions of the planets with mathematical precision, they believed them to be, like the stars, made of some celestial material that never decayed.

In ancient China, the whole universe, everything on earth and in the sky, was considered part of a giant organism. However, they envisaged a universe that was many millions of years old and, in that respect, they were close to today's view.

Western civilisation has grown up under the influence of Greek ideas and also that of Chinese teachings, which insist on a single God who is creator and sustainer of the universe. The Bible states that God created the universe but does not contain detailed scientific information about the process of the creation of the universe. Galileo used to be fond of saying that the Bible teaches the way to get to Heaven, not the way Heavens go. The Church, which was the dominant force in the land in those days, allowed no speculation into such divine matters. The misfortune of Copernicus, as a result of his statement that it was the sun and not the earth to be the centre of the spherical universe, is well known.

Even after the beginning of the modern scientific period, when Newton had worked out the motions of planets in great detail and also came up with the idea of universal gravitation, people still considered the creation of the universe to be a divine act that lies beyond speculation.

With the vast amount of information that was collected in the last century through the observation of deep space, as well as the development of the Relativity theory and Quantum Mechanics, scientists are at last in a position to work out how the universe began.

Today, speculation and scientific research into cosmology lies outside the realm of religion, to the extent that some scientists today do not consider the creation of the universe to be a divine act at all. However, the same scientists are well aware that when they trace the age of the universe to its origin, or to that moment that sparked the beginning, they too concede that science becomes unable to explain the events of that initial moment, for at that initial moment all the laws of physics seem to break down.

The failure of some scientists to contribute this initial moment of creation to a divine creator stems from the fact that they regard this initial moment as the moment when all the laws of physics break down, rather than being the moment when all the laws of physics came to be, or in other words, when all the laws of physics were set. The question which they do not seem to tackle is: by whom?

The creation of the universe is a subject that is given great attention in the Quran. The huge and varied amount of information contained in the Quran about almost every stage and aspect of the creation continues to astound scientists today because of its very accurate agreement with current knowledge. How can a book written in the 7th century contain such a rich amount of scientific information that was to be attained 14 centuries later? For those reasons, neutral and unbiased observers do consider this to be valid evidence that such a book could never have been the product of any human being.

All the evidence available today suggests an explosive origin to the universe, one that brought space, time and matter into existence. This is what is referred to as the Big Bang. The theory of the Big Bang which has successfully replaced the Steady State theory was worked out in the 1920's by two scientists quite independently of each other. They were the Russian meteorologist Alexksandr Friedmann and the Belgian mathematician Georges Lemaitre (Deep Space, Colin A. Ronan, p. 156).

The Big Bang itself resulted from an extremely dense singularity. The creation of the universe is one of matter, space and time that are intimately linked together. Matter and space were joined as one and then were separated in the explosion. This is very accurately described in the Quran:

> Have the disbelievers not seen that the skies (space) and the earth (matter) were joined together then We ripped them apart?
>
> **21:30**

The subsequent history of the Big Bang saw the Americans George Gamow, Ralph Alpher and Robert Horman indicate that the whole event took place at a very high temperature; it was a hot Big Bang. This view has been confirmed by the later discovery of the background microwave radiation. The eventual formation of galaxies resulted as a condensation, under gravitational pull, of hot gases which were mainly hydrogen, but may have also contained helium and a few other light elements. With the passing of time, and with the formation of galaxies, the gas gradually condensed into individual stars. The universe in its very early stages was, thus, still in the form of hot gases. This is confirmed in the Quran in the following verse:

Then He took hold of the sky while it was smoke.

41:11

Note that the verse did not say clouds or gas, but smoke, which is a very accurate description as smoke is hot gas, whilst clouds and gas could be hot or cold.

Once stars were formed, a system had to be devised to govern their motion. The kinetic energy stored in the forword movement of these bodies could not be relied upon on its own, otherwise stars and also planets would have shot off in straight lines dispersing into space. No planet would ever revolve around its mother star, which also applies to earth and thus, life would not have evolved on earth, because life on earth is so dependent on the heat, light and energy derived from the sun.

Gravity, a brilliant divine force, works as an equating factor to the centrifugal force to induce precise orbits for all heavenly bodies. The speed, mass and distance of two bodies have to be worked out very precisely to induce an orbit.

If you were to throw a tennis ball upwards towards the sky, it would travel upwards as a result of the kinetic energy stored in the throw but, eventually, the gravity of the earth will take over and the ball will fall back to the ground. But, if you were to throw the ball at a very high speed (say 10 km per second) it would escape the gravity of the earth and leave the earth altogether. This is what is known as the escape velocity. It is the speed required for a moving body to enable it to escape the gravity of a heavenly body such as a planet or a star.

When an artificial satellite is placed into orbit around the earth, what happens is that at a required distance, while the satellite is shooting out of the earth's gravitational field, its speed is reduced which reduces its kinetic energy and with some directional adjustments its kinetic energy can be equated with the earth's gravity. All these adjustments must be very precisely calculated and executed for an orbit to be obtained.

When one looks at the endless orbits of moons (around their planets), of planets (around their stars), of stars (around the centre of their galaxies), we can only gasp in awe at the intelligent force who designed all these orbits.

These very accurate balances are mentioned in the following verses:
The sun and the moon precisely calculated.

55:5

The following verse speaks of the balance which must be calculated to obtain the orbits:
And the sky, He raised it and set the balance.

55:7

The orbits of the heavenly bodies are mentioned in the verse:
And the sun and the moon, each swimming in an orbit.

21:33

The next stage sees these massive newly formed stars start to shrink under their own gravitational pull. As a result, their central regions become denser and thus, hotter. When the material in the centre of the star has heated up sufficiently, to be exact, at least seven million degrees K, nuclear reactions begin. These reactions, which are similar to those which take place in a hydrogen bomb, continue throughout the life of the star. These reactions are distinctly different from ordinary combustion of burning wood. What actually takes place inside a star is that hydrogen is converted to helium with the emission of huge energy.

The following verse speaks about brilliant stars and their fuel:
Its oil (fuel) almost lights up even though no fire has touched it.

24:35

The verse mentions a star, it's fuel and a reaction which is not combustion (fire). Short of saying 'nuclear reactions' the verse is a very accurate description of what goes on inside a star.

These nuclear reactions cause the stars to radiate all types of radiation into space, from x-rays and gamma rays in the short waves all the way to the longer radio waves. The visible section of those waves which are found between the ultra-violet and the infra-red is what we call sunlight.

On the other hand, the planets do not emit any light of their own, but only reflect light. This differentiation between natural light and reflected light is pointed out with the words:

Blessed is He who placed constellations in the sky and placed in it a lamp (sun), and a shining moon.

25:61

And also:

He is the One who rendered the sun to emit light and the moon that is lit.

10:5

In 1965, a very important discovery was made, and that was the background radiation which confirmed the Big Bang theory. The Big Bang theory, together with the detection of the red shift in the spectrum of far away galaxies, gave birth to yet the new concept of the expanding universe.

Further confirmation of the theory of the expanding universe was obtained from the spectrum analysis of far away galaxies. When you hear a police car or ambulance approaching you and then moving away, you will notice a change in the pitch of its siren. As the vehicle approaches, the siren wails at a higher pitch than when it moves away. Yet, in reality the siren is wailing at the same pitch all the time. To the driver of the vehicle the sound of the siren never changes. Why does this happen? The reason is that the waves of sound emitted by the siren change in frequency, which causes a change in pitch. This principle, which is called the Doppler Effect after its discoverer, applies to any waves and not only that of sound. When applied to light waves it was found that if the source of light is approaching its light would be shifted towards the blue end of the spectrum, while as light from a receding source would be shifted towards the red end of the spectrum. When analysing the light we receive from distant galaxies it was found that they all had a red shift meaning that they were flying away from us. This contribution of the red shift analysis meant that the universe is indeed expanding.

This conclusion is mentioned in the Quran:

We have constructed the universe with might (power) and We are expanding it.

51:47

Note that the word "expanding" is used in the present tense and not in the past which again is in agreement with the fact that the expansion of the universe is a continuous process.

When the Quran was revealed in the 7th century, it was still believed that all the stars in the sky, including our sun, were eternal and are made of a material that never fades or decays. No one was aware of the nature of the reactions that took place inside stars, for that was to be 20th century atomic theory territory. This information as shown was mentioned in the Quran.

Atomic reactions take place inside a star for a finite time, eventually running out of energy. When that stage is reached, a star like our sun will undergo a series of drastic changes. First, it will expand to become a red giant. The nearest planet, Mercury, would be swallowed up and the intense heat given off by the sun in this red giant stage would cause all the seas and oceans on earth to boil over and evaporate, signalling the end of life on earth. Eventually, the star would start to collapse and lose its lustre; it is then extinguished and ends up as a white dwarf. This stage of a star's life is described in the Quran with the words:

When the stars are extinguished.

77:8

The finite life of stars is also referred to:

He ordained the sun and the moon, each to run for a specified term.

13:2

Sura, 81 in its opening verses, describes the end of time as follows:

When the sun is rolled (swelling like a ball), and when the stars collapse.

81:1-2

When the seas are set aflame.

81:6

The age of the Universe

Whilst the Big Bang provided an explanation as to the origin of the universe, it still remained necessary to calculate its age. To do so, astronomers once again rely on red shift to calculate the speeds and distances of the furthest galaxies and quasars. These distances provide good indications to the age of the universe. The most distant quasars, which have velocities of some 240,000 km/sec (80% the speed of light), are at distances of up to 14 billion light years from Earth. When we look into the depth of space, we are actually looking back far into the past. When we are looking at such a distant quasar, we are not seeing it as it is now, but as it was 14 billion years ago.

The calculation of the age of the universe during the last 70 years or so has fluctuated between 10 and 20 billion years.

It is interesting to know that the age of the universe has been mentioned in the Quran. This information is found in the combined significance of the following two verses:

1. The angels and the Ruh ascend unto Him in a day, the measure of which was fifty thousand years.

70:4

This verse refers to the ascent of angels back to heaven after settling all matters of life in the universe.

The verse clearly said a day that "was" and not a day that "is", which clearly indicates that that day was in the past (50,000 years ago).

2. A day with your Lord is like a thousand years of your count.

22:47

With a few simple equations:

If 1 day (for God) = 1000 years (for man)

1 year (for God) = 1000 x 365 (for man)

= 365,000 years

Therefore, 50,000 years (for God) = 365,000 x 50,000 (for man)

= 18.25 billion!

The 50,000 years mentioned in point 1 above are of God's years and not of man's. This is because man was not mentioned at all in that verse, and more importantly because the subject of the verse (creation of the universe) is obviously a matter executed by God and not by man and so, its description is related to God and not to man.

This becomes evident when we compare this verse to other verses that clearly speak of years as related to man's count, such as:

A day which is equivalent to one thousand years of your count.

32:5

As we can see, the age of the universe given in the Quran (18:25 billion years) is older than the age agreed upon by scientists today which is 13.8 billion years.

However, when we look into the methods by which scientists arrived at the age of 13.8 billion years, we can understand why this age is no more than an estimated age. The 13.8 estimation could well be an under estimation for two reasons:

First:

The quasar that is estimated to be 14 billion light years away, and whoselight took 14 billion years to reach us does not necessarily mean that the universe is 14 billion years old for the following reason:

The light that reaches us from that quasar tells us that the quasar has been a quasar for at least 14 billion years. However, it does not tell us the length of time from the Big Bang up to the formation of that quasar! There must have been a period of time after the Big Bang when that quasar did not exist. How long was that period of time? To accurately calculate the age of the universe, this period of time must be added to the 14 billion years, which would mean that the universe is older than 14 billion years.

Second:

Astronomers estimate the age of the universe in two ways:
1. By looking for the oldest and farthest stars, galaxies and quasars.
2. By measuring the rate of expansion of the universe and extrapolating back to the Big Bang.

Both of these methods are based on available observations and are constantly revised with better apparatus and new discoveries.
1. The oldest stars: This method is based on the already detected stars in the observable universe. We can only detect what our instruments allow us to observe and detect. The observable universe has been consistently getter bigger with the advance in technology and instrumentation. There is no reason to think that this trend will stop in the future. A bigger universe, with farther stars and galaxies means an older universe.

2. The rate of expansion: Even though the rate of expansion of the universe is accelerating, it is dependent on what is called the Hubble constant, whose estimation has been increased a number of times. The Hubble constant is dependent on how much matter is in the universe, as well as the amount of dark energy in the universe. Both of these values are estimations at best.

Multiple Universes

When astronomers discuss the universe, they are always talking about the observable universe. If we go back to the days of the ancient Egyptians five thousand years ago, we find that their understanding of size of the universe was no more than the dome of the sky, covering the earth like the dome in a planetarium. The stars seemed, at the most, some thousands of kilometres away. The Greek astronomers, some two thousand or more years ago, thought of the universe as a sphere but still approximately of the same size.

Copernicus, who believed the sun and not the earth to be the centre of the spherical universe, thought it was much bigger than this, but not until about 170 years ago did anyone really know the distance of even the nearest stars.

Then they found that the distance to stars should be measured in millions of millions of kilometres. Even so, it was still a very small universe, with all the stars being together in one large star island.

It was not until the 1920's that astronomers discovered that our galaxy was only one of millions of others. Only then did astronomers start to appreciate the actual size of the universe.

Astronomers are today debating the issue of whether our universe is the only one in existence.

The idea of multiple universes is closely linked with the Black Hole concept. Certainly, the squashing of matter into an infinitely small area inside a Black Hole is in sharp disagreement with the Law of conservation of matter. It has been suggested that all matter falling into a Black Hole could be ejected into another time-space universe in what is referred to as a White Hole. Mathematical studies of space and time do show that this is possible in theory; but does it happen in practice? We do not know, but there certainly seems to be regions in deep space from where material is pouring out into our universe in what looks like a White Hole.

The jet of material from the active elliptical galaxy M 87 is a case in point. Has it come from a White Hole connected to a Black Hole in some other universe?

An alternate analysis that could also provide an equally valid justification for the existence of multiple universes is associated with the speed of light.

Between the years 1905–15, in his Theory of Relativity, Albert Einstein stated that the speed of light is a limiting velocity in the universe; nothing can travel faster than light. His theory also outlined that the speed of light is constant, unaffected by the movement of its source and independent of all observers. Quasars, which are the most distant objects in the universe travel at speeds approaching 80% of the speed of light but nothing travels faster than the speed of light.

Could it be that the speed of light acts as a gate, a valve or a barrier between our universe and other universes? Perhaps a different time-space universe where matter, if it can still be called so, exists and is travelling at speeds faster than that of light? We cannot cross that barrier ourselves nor can any physical matter, but there are strong indications that there is some kind of existence on the other side.

A mention should be given here to some speculation concerning particles called tachyons which occur in some nuclear reactions. These particles are believed to travel faster than light. They can never travel at the speed of light, only faster. In short, tachyons behave in just the opposite way from matter, but as no one has actually observed a tachyon, could the reason be because they exist in a different dimension of time-space?

When we examine the verses in the Quran that relate to this subject, not only do we find incredible information concerning the creation and existence of multiple universes, but also regarding the barriers that lie between them.

Multiple universes are mentioned in more than one verse in the Quran such as:

Did you not see that God created seven universes in layers?

71:15

The barriers that exist between these universes are mentioned in the following verse:

If you are able to penetrate through the regions of the heavens and the earth, then go ahead and penetrate; you cannot penetrate without authority.

55:33

The word "penetrate" implies the existence of barriers between the zones of the skies (universes).

Our universe, as we know it today, includes within it all the stars and galaxies that we have detected in the sky so far. If multiple universes exist, they lie outside and beyond all these galaxies. This too seems to be in harmony with the information given in the Quran:

We adorned the lowest universe with lamps (stars).

41:12

Some interpreters of the Quran suggested that the earth's atmosphere is the first universe, that the region beyond the earth's atmosphere up to the moon forms the second universe, while the solar system constitutes the third universe and so on. This interpretation is clearly inaccurate since it contradicts the words in 41:12, which tell us that God adorned the "lowest" universe with lamps (stars). Clearly there are no stars in the earth's atmosphere! The correct meaning of 41:12 is that all the observable universe, with all its stars and galaxies, constitute the lowest universe.

CHAPTER 17

BIG BANG

From Wikipedia, the free encyclopedia
Jump to: navigation, search
"Big Bang theory" redirects here. For the American TV sitcom, see The Big Bang Theory. For other uses, see Big Bang (disambiguation) and Big Bang Theory (disambiguation).

The **Big Bang** theory is the prevailing cosmological model for the universe from the earliest known periods through its subsequent large-scale evolution.[114,115,116] The model describes how the universe expanded from a very high density and high temperature state,[117,118] and offers a comprehensive explanation for a broad range of phenomena, including the abundance of light elements, the cosmic microwave background, large scale structure and Hubble's Law.[119] If the known laws of physics

---

[114] Joseph Silk (2009). *Horizons of Cosmology*. Templeton Press. p. 208.
[115] Simon Singh (2005). *Big Bang: The Origin of the Universe*. Harper Perennial. p. 560.
[116] Wollack, E. J. (10 December 2010). "Cosmology: The Study of the Universe." *Universe 101: Big Bang Theory*. NASA. Archived from the original on 14 May 2011. Retrieved 27 April 2011. The second section discusses the classic tests of the Big Bang theory that make it so compelling as the likely valid description of our universe.
[117] a b c d e "First Second of the Big Bang." *How The Universe Works 3*. 2014. Discovery Science.
[118] "Big-bang model." *Encyclopædia Britannica*. Retrieved 11 February 2015.
[119] a b Wright, E. L. (9 May 2009). "What is the evidence for the Big Bang?." *Frequently Asked Questions in Cosmology*. UCLA, Division of Astronomy and Astrophysics. Retrieved 16 October 2009.

are extrapolated to the highest density regime, the result is a singularity which is typically associated with the Big Bang. Detailed measurements of the expansion rate of the universe place this moment at approximately 13.8 billion years ago, which is thus considered the age of the universe. After the initial expansion, the universe cooled sufficiently to allow the formation of subatomic particles, and later simple atoms. Giant clouds of these primordial elements later coalesced through gravity in halos of dark matter, eventually forming the stars and galaxies visible today.

Since Georges Lemaître first noted in 1927 that an expanding universe could be traced back in time to an originating single point, scientists have built on his idea of cosmic expansion. While the scientific community was once divided between supporters of two different expanding universe theories, the Big Bang and the Steady State theory, empirical evidence provides strong support for the former. In 1929, from analysis of galactic redshifts, Edwin Hubble concluded that galaxies are drifting apart; this is important observational evidence consistent with the hypothesis of an expanding universe. In 1964 the cosmic microwave background radiation was discovered, which was crucial evidence in favor of the Big Bang model, since that theory predicted the existence of background radiation throughout the universe before it was discovered. More recently, measurements of the redshifts of supernovae indicate that the expansion of the universe is accelerating, an observation attributed to dark energy's existence. The known physical laws of nature can be used to calculate the characteristics of the universe in detail back in time to an initial state of extreme density and temperature.

Overview

American astronomer Edwin Hubble observed that the distances to faraway galaxies were strongly correlated with their **redshifts**. This was interpreted to mean that all distant galaxies and clusters are receding away from our vantage point with an apparent velocity proportional to their distance: that is, the farther they are, the faster they move away from us, regardless of direction. Assuming the **Copernican principle** (that the Earth is not the center of the universe), the only remaining interpretation is that all observable regions of the universe are receding from all others. Since we know that the distance between galaxies increases today, it must mean that in the past galaxies were closer together. The continuous expansion of the universe implies that the universe was denser and hotter in the past.

Large **particle accelerators** can replicate the conditions that prevailed after the early moments of the universe, resulting in confirmation and refinement of the details of the Big Bang model. However, these accelerators can only probe so far into **high energy regimes**. Consequently, the state of the universe in the earliest instants of the Big Bang expansion is still poorly understood and an area of open investigation and speculation.

The first subatomic particles to be formed included protons, neutrons, and electrons. Though simple atomic nuclei formed within the first three minutes after the Big Bang, thousands of years passed before the first electrically neutral atoms formed. The majority of atoms produced by the Big Bang were hydrogen, along with helium and traces of lithium. Giant clouds of these primordial elements later coalesced through gravity to form stars and galaxies, and the heavier elements were synthesized either within stars or during supernovae.

The Big Bang theory offers a comprehensive explanation for a broad range of observed phenomena, including the abundance of light elements, the **cosmic microwave background**, **large scale structure**, and **Hubble's Law**.[6] The framework for the Big Bang model relies on **Albert Einstein**'s theory of **general relativity** and on simplifying assumptions such as **homogeneity** and **isotropy** of space. The governing equations were formulated by **Alexander Friedmann**, and similar solutions were worked on by **Willem de Sitter**. Since then, astrophysicists have incorporated observational and theoretical additions into the Big Bang model, and its **parametrization** as the **Lambda-CDM model** serves as the framework for current investigations of theoretical cosmology. The Lambda-CDM model is the current "standard model" of Big Bang cosmology, **consensus** is that it is the simplest model that can account for the various measurements and **observations relevant to cosmology**.

Timeline

Main article: Chronology of the universe

Singularity

See also: Gravitational singularity and Planck epoch

Extrapolation of the expansion of the universe backwards in time using general relativity yields an infinite density and temperature at a finite time in the past.[13] **This singularity indicates that general relativity is not an adequate description of the laws of physics in this regime. How closely models based on general relativity alone can be used to extrapolate toward the singularity is debated—certainly no closer than the end of the Planck epoch.**

This primordial singularity is itself sometimes called "the Big Bang", but the term can also refer to a more generic early hot, dense phase of the universe. In either case, "the Big Bang" as an event is also colloquially referred to as the "birth" of our universe since it represents the point in history where the universe can be verified to have entered into a **regime** where the laws of physics as we understand them (specifically general relativity and the **standard model of particle physics**) work. Based on measurements of the expansion using **Type Ia supernovae** and measurements of temperature fluctuations in the **cosmic microwave background**, the time that has passed since that event, otherwise known as the "**age of the universe**" is 13.799 ± 0.021 billion years.[16] The agreement of independent measurements of this age supports the **ΛCDM model** that describes in detail the characteristics of the universe.

Inflation and baryogenesis

Main articles: Cosmic inflation and baryogenesis

The earliest phases of the Big Bang are subject to much speculation. In the most common models the universe was filled **homogeneously** and **isotropically** with a very high **energy density** and huge temperatures and **pressures** and was very rapidly expanding and cooling. Approximately 10^{-37} seconds into the expansion, a **phase transition** caused a **cosmic inflation**, during which the universe grew **exponentially** during which time **density fluctuations** that occurred because of the **uncertainty principle** were amplified into the seeds that would later form the **large-scale structure** of the universe. After inflation stopped, **reheating** occurred until the universe obtained the temperatures required for the **production** of a **quark–gluon plasma** as well as all other **elementary particles**. Temperatures were so high that the random motions of particles were at **relativistic speeds**, and **particle–antiparticle pairs** of all kinds were being continuously created and destroyed in collisions. [4] At some point, an unknown reaction called **baryogenesis** violated the conservation of **baryon number**, leading to a very small excess of **quarks** and **leptons** over antiquarks and antileptons—of the order of one part in 30 million. This resulted in the predominance of **matter** over **antimatter** in the present universe.

Cooling

Main articles: Big Bang nucleosynthesis and cosmic microwave background radiation Panoramic view of the entire near-infrared sky reveals the distribution of galaxies beyond the Milky Way. Galaxies are color-coded by redshift.

The universe continued to decrease in density and fall in temperature, hence the typical energy of each particle was decreasing. **Symmetry breaking** phase transitions put the **fundamental forces** of physics and the parameters of **elementary particles** into their present form. After about 10^{-11} seconds, the picture becomes less speculative, since particle energies drop to values that can be attained in **particle accelerators**. At about 10^{-6} seconds, quarks and gluons combined to form **baryons** such as protons and neutrons. The small excess of quarks over antiquarks led to a small excess of baryons over antibaryons. The temperature was now no longer high enough to create new proton–antiproton pairs (similarly for neutrons–antineutrons), so a mass annihilation immediately followed, leaving just one in 10^{10} of the original protons and neutrons, and none of their antiparticles. A similar process happened at about 1 second for electrons and positrons. After these annihilations, the remaining protons, neutrons and electrons were no longer moving relativistically and the energy density of the universe was dominated by **photons** (with a minor contribution from **neutrinos**).

A few minutes into the expansion, when the temperature was about a billion (one thousand million) **kelvin** and the density was about that of air, neutrons combined with protons to form the universe's **deuterium** and helium **nuclei** in a process called **Big Bang nucleosynthesis**. Most protons remained uncombined as hydrogen nuclei. As the universe cooled, the **rest mass** energy density of matter came to gravitationally dominate that of the photon **radiation**. After about 379,000 years the electrons and nuclei combined into atoms (mostly hydrogen); hence the radiation decoupled from matter and continued through space largely unimpeded. This relic radiation is known as the **cosmic microwave background radiation**. The **chemistry of life** may have begun shortly after the Big Bang, **13.8 billion years ago**, during a habitable epoch when the universe was only 10–17 million years old.

Structure formation

Main article: Structure formation

Abell 2744 galaxy cluster – Hubble Frontier Fields view.

Over a long period of time, the slightly denser regions of the nearly uniformly distributed matter gravitationally attracted nearby matter and thus grew even denser, forming gas clouds, stars, galaxies, and the other astronomical structures observable today.[4] The details of this process depend on the amount and type of matter in the universe. The four possible types of matter are known as **cold dark matter, warm dark matter, hot dark matter,** and **baryonic matter.** The best measurements available (from **WMAP**) show that the data is well-fit by a Lambda-CDM model in which dark matter is assumed to be cold (warm dark matter is ruled out by early **reionization**), and is estimated to make up about 23% of the matter/energy of the universe, while baryonic matter makes up about 4.6%. In an "extended model" which includes hot dark matter in the form of **neutrinos**, then if the "physical baryon density" $\Omega_b h^2$ is estimated at about 0.023 (this is different from the 'baryon density' Ω_b expressed as a fraction of the total matter/energy density, which as noted above is about 0.046), and the corresponding cold dark matter density $\Omega_c h^2$ is about 0.11, the corresponding neutrino density $\Omega_v h^2$ is estimated to be less than 0.0062.

Cosmic acceleration

Main article: Accelerating universe

Independent lines of evidence from Type Ia supernovae and the **CMB** imply that the universe today is dominated by a mysterious form of energy known as **dark energy**, which apparently permeates all of space. The observations suggest 73% of the total energy density of today's universe is in this form. When the universe was very young, it was likely infused with dark energy, but with less space and everything closer together, gravity predominated, and it was slowly braking the expansion. But eventually, after numerous billion years of expansion, the growing abundance of dark energy caused the **expansion of the universe** to slowly begin to accelerate. Dark energy in its simplest formulation takes the form of the **cosmological constant** term in **Einstein's field equations** of general relativity, but its composition and mechanism are unknown and, more generally, the details of its **equation of state** and relationship with the **Standard Model** of particle physics continue to be investigated both through observation and theoretically.

All of this cosmic evolution after the **inflationary epoch** can be rigorously described and modeled by the ΛCDM model of cosmology, which uses the independent frameworks of quantum mechanics and Einstein's General Relativity. There is no well-supported model describing the action prior to 10^{-15} seconds or so. Apparently a new unified theory of **quantum gravitation** is needed to break this barrier. Understanding this earliest of eras in the history of the universe is currently one of the greatest **unsolved problems in physics**.

Features of the model

The Big Bang theory depends on two major assumptions: the universality of physical laws and the cosmological principle. The cosmological principle states that on large scales the universe is homogeneous and isotropic.

These ideas were initially taken as postulates, but today there are efforts to test each of them. For example, the first assumption has been tested by observations showing that largest possible deviation of the **fine structure constant** over much of the **age of the universe** is of order 10^{-5}. Also, general relativity has passed stringent **tests** on the scale of the Solar System and binary stars.

If the large-scale universe appears isotropic as viewed from Earth, the cosmological principle can be derived from the simpler **Copernican principle**, which states that there is no preferred (or special) observer or vantage point. To this end, the cosmological principle has been confirmed to a level of 10^{-5} via observations of the CMB. The universe has been measured to be homogeneous on the largest scales at the 10% level.

Expansion of space

Main articles: Friedmann–Lemaître–Robertson–Walker metric and Metric expansion of space

General relativity describes spacetime by a **metric**, which determines the distances that separate nearby points. The points, which can be galaxies, stars, or other objects, themselves are specified using a **coordinate chart** or "grid" that is laid down over all **spacetime**. The cosmological principle implies that the metric should be **homogeneous** and **isotropic** on large scales, which uniquely singles out the **Friedmann–Lemaître–Robertson–Walker metric** (FLRW metric). This metric contains a **scale factor**, which describes how the size of the universe changes with time. This enables a convenient choice of a **coordinate system** to be made, called **comoving coordinates**. In this coordinate system the grid expands along with the universe, and objects that are moving only because of the expansion of the universe remain at fixed points on the grid. While their *coordinate* distance (**comoving distance**) remains constant, the *physical* distance between two such comoving points expands proportionally with the **scale factor** of the universe.

The Big Bang is not an explosion of matter moving outward to fill an empty universe. Instead, **space itself expands** with time everywhere and increases the physical distance between two comoving points. In other words, the Big Bang is not an explosion *in space*, but rather an expansion *of space*.[4] Because the FLRW metric assumes a uniform distribution of mass and energy, it applies to our universe only on large scales—local concentrations of matter such as our galaxy are gravitationally bound and as such do not experience the large-scale expansion of space.

Main article: Cosmological horizon.

An important feature of the Big Bang spacetime is the presence of horizons. Since the universe has a finite age, and light travels at a finite speed, there may be events in the past whose light has not had time to reach us. This places a limit or a *past horizon* on the most distant objects that can be observed. Conversely, because space is expanding, and more distant objects are receding ever more quickly, light emitted by us today may never "catch up" to very distant objects. This defines a *future horizon*, which limits the events in the future that we will be able to influence. The presence of either type of horizon depends on the details of the FLRW model that describes our universe. Our understanding of the universe back to very early times suggests that there is a past horizon, though in practice our view is also limited by the opacity of the universe at early times. So our view cannot extend further backward in time, though the horizon recedes in space. If the expansion of the universe continues to accelerate, there is a future horizon as well.

Main article: History of the Big Bang theory

See also: Timeline of cosmology

Etymology

English astronomer **Fred Hoyle** is credited with coining the term "Big Bang" during a 1949 BBC radio broadcast. It is popularly reported that Hoyle, who favored an alternative **"steady state"** cosmological model, intended this to be pejorative, but Hoyle explicitly denied this and said it was just a striking image meant to highlight the difference between the two models.

Development

Hubble eXtreme Deep Field (XDF)

XDF size compared to the size of the Moon – several thousand galaxies, each consisting of billions of stars, are in this small view.

XDF (2012) view – each light speck is a galaxy – some of these are as old as 13.2 billion years – the universe is estimated to contain 200 billion galaxies.

XDF image shows fully mature galaxies in the foreground plane – nearly mature galaxies from 5 to 9 billion years ago – protogalaxies, blazing with young stars, beyond 9 billion years.

The Big Bang theory developed from observations of the structure of the universe and from theoretical considerations. In 1912 Vesto

Slipher measured the first Doppler shift of a "spiral nebula" (spiral nebula is the obsolete term for spiral galaxies), and soon discovered that almost all such nebulae were receding from Earth. He did not grasp the cosmological implications of this fact, and indeed at the time it was highly controversial whether or not these nebulae were "island universes" outside our Milky Way. Ten years later, Alexander Friedmann, a Russian cosmologist and mathematician, derived the Friedmann equations from Albert Einstein's equations of general relativity, showing that the universe might be expanding in contrast to the static universe model advocated by Einstein at that time. In 1924 Edwin Hubble's measurement of the great distance to the nearest spiral nebulae showed that these systems were indeed other galaxies. Independently deriving Friedmann's equations in 1927, Georges Lemaître, a Belgian physicist and Roman Catholic priest, proposed that the inferred recession of the nebulae was due to the expansion of the universe.

In 1931 Lemaître went further and suggested that the evident expansion of the universe, if projected back in time, meant that the further in the past the smaller the universe was, until at some finite time in the past all the mass of the universe was concentrated into a single point, a "primeval atom" where and when the fabric of time and space came into existence.

Starting in 1924, Hubble painstakingly developed a series of distance indicators, the forerunner of the cosmic distance ladder, using the 100-inch (2.5 m) Hooker telescope at Mount Wilson Observatory. This allowed him to estimate distances to galaxies whose redshifts had already been measured, mostly by Slipher. In 1929 Hubble discovered a correlation between distance and recession velocity—now known as Hubble's law. Lemaître had already shown that this was expected, given the cosmological principle.

In the 1920s and 1930s almost every major cosmologist preferred an eternal steady state universe, and several complained that the beginning of time implied by the Big Bang imported religious concepts into physics;

this objection was later repeated by supporters of the steady state theory. This perception was enhanced by the fact that the originator of the Big Bang theory, Monsignor Georges Lemaître, was a Roman Catholic priest. Arthur Eddington agreed with Aristotle that the universe did not have a beginning in time, viz., that matter is eternal. A beginning in time was "repugnant" to him. Lemaître, however, thought that:

If the world has begun with a single quantum, the notions of space and time would altogether fail to have any meaning at the beginning; they would only begin to have a sensible meaning when the original quantum had been divided into a sufficient number of quanta. If this suggestion is correct, the beginning of the world happened a little before the beginning of space and time.

During the 1930s other ideas were proposed as non-standard cosmologies to explain Hubble's observations, including the Milne model, the oscillatory universe (originally suggested by Friedmann, but advocated by Albert Einstein and Richard Tolman) and Fritz Zwicky's tired light hypothesis.

After World War II, two distinct possibilities emerged. One was Fred Hoyle's steady state model, whereby new matter would be created as the universe seemed to expand. In this model the universe is roughly the same at any point in time. The other was Lemaître's Big Bang theory, advocated and developed by George Gamow, who introduced big bang nucleosynthesis (BBN) and whose associates, Ralph Alpher and Robert Herman, predicted the cosmic microwave background radiation (CMB). Ironically, it was Hoyle who coined the phrase that came to be applied to Lemaître's theory, referring to it as "this *big bang* idea" during a BBC Radio broadcast in March 1949. For a while, support was split between these two theories. Eventually, the observational evidence, most notably from radio source counts, began to favor Big Bang over Steady State. The discovery and confirmation of the cosmic microwave background radiation in 1964 secured the Big Bang as the best theory of the origin and evolution of the universe. Much of the current work in cosmology includes understanding how galaxies form in the context of the Big Bang, understanding the physics of the universe at earlier and earlier times, and reconciling observations with the basic theory.

In 1968 and 1970 Roger Penrose, Stephen Hawking, and George F. R. Ellis published papers where they showed that mathematical singularities were an inevitable initial condition of general relativistic models of the Big Bang. Then, from the 1970s to the 1990s, cosmologists worked on characterizing the features of the Big Bang universe and resolving outstanding problems. In 1981 Alan Guth made a breakthrough in theoretical work on resolving certain outstanding theoretical problems in the Big Bang theory with the introduction of an epoch of rapid expansion in the early universe he called "inflation." Meanwhile, during these decades, two questions in observational cosmology that generated much discussion and disagreement were over the precise values of the Hubble Constant and the matter-density of the universe (before the discovery of dark energy, thought to the key predictor for the eventual fate of the universe). In the mid-1990s observations of certain globular clusters appeared to indicate that they were about 15 billion years old, which conflicted with most then-current estimates of the age of the universe (and indeed with the age measured today). This issue was later resolved when new computer simulations, which included the effects of mass loss due to stellar winds, indicated a much younger age for globular clusters. While there still remain some questions as to how accurately the ages of the clusters are measured, globular clusters are of interest to cosmology as some of the oldest objects in the universe.

Significant progress in Big Bang cosmology have been made since the late 1990s as a result of advances in telescope technology as well as the analysis of data from satellites such as COBE, the Hubble Space Telescope and WMAP. Cosmologists now have fairly precise and accurate measurements of many of the parameters of the Big Bang model, and have made the unexpected discovery that the expansion of the universe appears to be accelerating.

Observational evidence

Artist's depiction of the WMAP satellite gathering data to help scientists understand the Big Bang "[The] big bang picture is too firmly grounded in data from every area to be proved invalid in its general features."

Lawrence Krauss

The earliest and most direct observational evidence of the validity of the theory are the expansion of the universe according to Hubble's law (as indicated by the redshifts of galaxies), discovery and measurement of the cosmic microwave background and the relative abundances of light elements produced by Big Bang nucleosynthesis. More recent evidence includes observations of galaxy formation and evolution, and the distribution of large-scale cosmic structures. These are sometimes called the "four pillars" of the Big Bang theory.

Precise modern models of the Big Bang appeal to various exotic physical phenomena that have not been observed in terrestrial laboratory experiments or incorporated into the Standard Model of particle physics. Of these features, dark matter is currently subjected to the most active laboratoryinvestigations. Remaining issues include the cuspy halo problem and the dwarf galaxy problem of cold dark matter. Dark energy is also an area of intense interest for scientists, but it is not clear whether direct detection of dark energy will be possible. Inflation and baryogenesis remain more speculative features of current Big Bang models. Viable, quantitative explanations for such phenomena are still being sought. These are currently unsolved problems in physics.

Hubble's law and the expansion of space

Main articles: Hubble's law and Metric expansion of space

See also: Distance measures (cosmology) and Scale factor (universe)

Observations of distant galaxies and quasars show that these objects are redshifted—the light emitted from them has been shifted to longer wavelengths. This can be seen by taking a frequency spectrum of an object and matching the spectroscopic pattern of emission lines or absorption lines corresponding to atoms of the chemical elements interacting with the light. These redshifts are uniformly isotropic, distributed evenly among the observed objects in all directions. If the redshift is interpreted as a Doppler shift, the recessional velocity of the object can be calculated. For some galaxies, it is possible to estimate distances via the cosmic distance ladder. When the recessional velocities are plotted against these distances, a linear relationship known as Hubble's law is observed:

$v = H_0 D$,

where

- v is the recessional velocity of the galaxy or other distant object,
- D is the comoving distance to the object, and
- H_0 is Hubble's constant, measured to be 70.4+1.3 −1.4 km/s/Mpc by the WMAP probe.

Hubble's law has two possible explanations. Either we are at the center of an explosion of galaxies—which is untenable given the Copernican principle—or the universe is uniformly expanding everywhere. This universal expansion was predicted from general relativity by Alexander Friedmann in 1922 and Georges Lemaître in 1927, well before Hubble made his 1929 analysis and observations, and it remains the cornerstone of the Big Bang theory as developed by Friedmann, Lemaître, Robertson, and Walker.

The theory requires the relation $v = HD$ to hold at all times, where D is the comoving distance, v is the recessional velocity, and v, H, and D vary as the universe expands (hence we write H_0 to denote the present-day Hubble "constant"). For distances much smaller than the size of the observable universe, the Hubble redshift can be thought of as the Doppler shift corresponding to the recession velocity v. However, the redshift is not a true Doppler shift, but rather the result of the expansion of the universe between the time the light was emitted and the time that it was detected.

That space is undergoing metric expansion is shown by direct observational evidence of the Cosmological principle and the Copernican principle, which together with Hubble's law have no other explanation. Astronomical redshifts are extremely isotropic and homogeneous, supporting the Cosmological principle that the universe looks the same in all directions, along with much other evidence. If the redshifts were the result of an explosion from a center distant from us, they would not be so similar in different directions.

Measurements of the effects of the cosmic microwave background radiation on the dynamics of distant astrophysical systems in 2000 proved the Copernican principle, that, on a cosmological scale, the Earth is not in a central position. Radiation from the Big Bang was demonstrably warmer at earlier times throughout the universe. Uniform cooling of the cosmic microwave background over billions of years is explainable only if the universe is experiencing a metric expansion, and excludes the possibility that we are near the unique center of an explosion.

Cosmic microwave background radiation

Main article: Cosmic microwave background radiation

9 year WMAP image of the cosmic microwave background radiation (2012). The radiation is isotropic to roughly one part in 100,000.

In 1964 Arno Penzias and Robert Wilson serendipitously discovered the cosmic background radiation, an omnidirectional signal in the microwave band. Their discovery provided substantial confirmation of the big-bang predictions by Alpher, Herman and Gamow around 1950. Through the 1970s the radiation was found to be approximately consistent with a black body spectrum in all directions; this spectrum has been redshifted by the expansion of the universe, and today corresponds to approximately 2.725 K. This tipped the balance of evidence in favor of the Big Bang model, and Penzias and Wilson were awarded a Nobel Prize in 1978.

The cosmic microwave background spectrum measured by the FIRAS instrument on the COBE satellite is the most-precisely measured black body spectrum in nature. The data points and error bars on this graph are obscured by the theoretical curve.

The *surface of last scattering* corresponding to emission of the CMB occurs shortly after *recombination*, the epoch when neutral hydrogen becomes stable. Prior to this, the universe comprised a hot dense photon-baryon plasma sea where photons were quickly scattered from free charged particles. Peaking at around 372±14 kyr, the mean free path for a photon becomes long enough to reach the present day and the universe becomes transparent.

In 1989 NASA launched the Cosmic Background Explorer satellite (COBE) which made two major advances: in 1990, high-precision spectrum measurements showed the CMB frequency spectrum is an almost perfect blackbody with no deviations at a level of 1 part in 10^4, and measured a residual temperature of 2.726 K (more recent measurements have revised this figure down slightly to 2.7255 K); then in 1992 further COBE measurements discovered tiny fluctuations (anisotropies) in the CMB temperature across the sky, at a level of about one part in 10^5. John C. Mather and George Smoot were awarded the 2006 Nobel Prize in Physics for their leadership in these results. During the following decade, CMB anisotropies were further investigated by a large number of ground-based and balloon experiments. In 2000–2001 several experiments, most notably BOOMERanG, found the shape of the universe to be spatially almost flat by measuring the typical angular size (the size on the sky) of the anisotropies.

In early 2003 the first results of the Wilkinson Microwave Anisotropy Probe (WMAP) were released, yielding what were at the time the most accurate values for some of the cosmological parameters. The results disproved several specific cosmic inflation models, but are consistent with the inflation theory in general. The Planck space probe was launched in May 2009. Other ground and balloon based cosmic microwave background experiments are ongoing.

Abundance of primordial elements

Main article: Big Bang nucleosynthesis

Using the Big Bang model it is possible to calculate the concentration of helium-4, helium-3, deuterium, and lithium-7 in the universe as ratios to the amount of ordinary hydrogen. The relative abundances depend on a single parameter, the ratio of photons to baryons. This value can be calculated independently from the detailed structure of CMB fluctuations. The ratios predicted (by mass, not by number) are about 0.25 for ^{4}He/H, about 10^{-3} for ^{2}H/H, about 10^{-4} for ^{3}He/H and about 10^{-9} for ^{7}Li/H.

The measured abundances all agree at least roughly with those predicted from a single value of the baryon-to-photon ratio. The agreement is excellent for deuterium, close but formally discrepant for ^{4}He, and off by a factor of two for ^{7}Li; in the latter two cases there are substantial systematic uncertainties. Nonetheless, the general consistency with abundances predicted by Big Bang nucleosynthesis is strong evidence for the Big Bang, as the theory is the only known explanation for the relative abundances of light elements, and it is virtually impossible to "tune" the Big Bang to produce much more or less than 20–30% helium. Indeed, there is no obvious reason outside of the Big Bang that, for example, the young universe (i.e., before star formation, as determined by studying matter supposedly free of stellar nucleosynthesis products) should have more helium than deuterium or more deuterium than ^{3}He, and in constant ratios, too.

Galactic evolution and distribution

Main articles: Galaxy formation and evolution and Structure formation

Detailed observations of the morphology and distribution of galaxies and quasars are in agreement with the current state of the Big Bang theory. A combination of observations and theory suggest that the first quasars and galaxies formed about a billion years after the Big Bang, and since then larger structures have been forming, such as galaxy clusters and superclusters. Populations of stars have been aging and evolving, so that distant galaxies (which are observed as they were in the early universe) appear very different from nearby galaxies (observed in a more recent state). Moreover, galaxies that formed relatively recently appear markedly different from galaxies formed at similar distances but shortly after the Big Bang. These observations are strong arguments against the steady-state model. Observations of star formation, galaxy and quasar distributions and larger structures agree well with Big Bang simulations of the formation of structure in the universe and are helping to complete details of the theory.

Primordial gas clouds

Focal plane of BICEP2 telescope under a microscope – used to search for polarization in the CMB.

In 2011 astronomers found what they believe to be pristine clouds of primordial gas, by analyzing absorption lines in the spectra of distant quasars. Before this discovery, all other astronomical objects have been observed to contain heavy elements that are formed in stars. These two clouds of gas contain no elements heavier than hydrogen and deuterium. Since the clouds of gas have no heavy elements, they likely formed in the first few minutes after the Big Bang, during Big Bang nucleosynthesis.

Other lines of evidence

The age of the universe as estimated from the Hubble expansion and the CMB is now in good agreement with other estimates using the ages of the oldest stars, both as measured by applying the theory of stellar evolution to globular clusters and through radiometric dating of individual Population II stars.

The prediction that the CMB temperature was higher in the past has been experimentally supported by observations of very low temperature absorption lines in gas clouds at high redshift. This prediction also implies that the amplitude of the Sunyaev–Zel'dovich effect in clusters of galaxies does not depend directly on redshift. Observations have found this to be roughly true, but this effect depends on cluster properties that do change with cosmic time, making precise measurements difficult.

Future observations

Future gravitational waves observatories might see primordial gravitational waves, relics of the early universe, up to less than a second of the Big Bang.

Problems and related issues in physics

See also: List of unsolved problems in physics

As with any theory, a number of mysteries and problems have arisen as a result of the development of the Big Bang theory. Some of these mysteries and problems have been resolved while others are still outstanding. Proposed solutions to some of the problems in the Big Bang model have revealed new mysteries of their own. For example, the **horizon problem**, the **magnetic monopole problem**, and the **flatness problem** are most commonly resolved with **inflationary theory**, but the details of the inflationary universe are still left unresolved and many, including some founders of the theory, say it has been disproven. What follows are a list of the mysterious aspects of the Big Bang theory still under intense investigation by cosmologists and astrophysicists.

Baryon asymmetry

Main article: Baryon asymmetry

It is not yet understood why the universe has more matter than antimatter. It is generally assumed that when the universe was young and very hot, it was in statistical equilibrium and contained equal numbers of baryons and antibaryons. However, observations suggest that the universe, including its most distant parts, is made almost entirely of matter. A process called baryogenesis was hypothesized to account for the asymmetry. For baryogenesis to occur, the **Sakharov conditions** must be satisfied. These require that baryon number is not conserved, that **C-symmetry** and **CP-symmetry** are violated and that the universe depart from **thermodynamic equilibrium**. All these conditions occur in the **Standard Model**, but the effects are not strong enough to explain the present baryon asymmetry.

Dark energy

Main article: Dark energy

Measurements of the redshift–magnitude relation for type Ia supernovae indicate that the expansion of the universe has been accelerating since the universe was about half its present age. To explain this acceleration, general relativity requires that much of the energy in the universe consists of a component with large negative pressure, dubbed "dark energy." Dark energy, though speculative, solves numerous problems. Measurements of the cosmic microwave background indicate that the universe is very nearly spatially flat, and therefore according to general relativity the universe must have almost exactly the critical density of mass/energy. But the mass density of the universe can be measured from its gravitational clustering, and is found to have only about 30% of the critical density.[10] Since theory suggests that dark energy does not cluster in the usual way it is the best explanation for the "missing" energy density. Dark energy also helps to explain two geometrical measures of the overall curvature of the universe, one using the frequency of gravitational lenses, and the other using the characteristic pattern of the large-scale structure as a cosmic ruler.

Negative pressure is believed to be a property of **vacuum energy**, but the exact nature and existence of dark energy remains one of the great mysteries of the Big Bang. Results from the WMAP team in 2008 are in accordance with a universe that consists of 73% dark energy, 23% dark matter, 4.6% regular matter and less than 1% neutrinos. According to theory, the energy density in matter decreases with the expansion of the universe, but the dark energy density remains constant (or nearly so) as the universe expands. Therefore, matter made up a larger fraction of the total energy of the universe in the past than it does today, but its fractional contribution will fall in the **far future** as dark energy becomes even more dominant.

The dark energy component of the universe has been explained by theorists using a variety of competing theories including Einstein's **cosmological constant** but also extending to more exotic forms of **quintessence** or other modified gravity schemes. A **cosmological constant problem** sometimes called the "most embarrassing problem in physics" results from the apparent discrepancy between the measured energy density of dark energy and the one naively predicted from **Planck units**.

Dark matter

Main article: Dark matter

Chart shows the proportion of different components of the Universe – about 95% is dark matter and dark energy.

During the 1970s and 80s, various observations showed that there is not sufficient visible matter in the universe to account for the apparent strength of gravitational forces within and between galaxies. This led to the idea that up to 90% of the matter in the universe is dark matter that does not emit light or interact with normal **baryonic** matter. In addition, the assumption that the universe is mostly normal matter led to predictions that was strongly inconsistent with observations.

In particular, the universe today is far lumpier and contains far less deuterium than can be accounted for without dark matter. While dark matter has always been controversial, it is inferred by various observations: the anisotropies in the CMB, **galaxy cluster** velocity dispersions, large-scale structure distributions, **gravitational lensing** studies, and **X-ray measurements** of galaxy clusters.

Indirect evidence for dark matter comes from its gravitational influence on other matter, as no dark matter particles have been observed in laboratories. Many **particle physics** candidates for dark matter have been proposed, and several projects to detect them directly are underway.

Additionally, there are outstanding problems associated with the currently favored cold dark matter model which include the dwarf galaxy problem and the cuspy halo problem. Alternative theories have been proposed that do not require a large amount of undetected matter but instead modify the laws of gravity established by Newton and Einstein, but no alternative theory has been as successful as the cold dark matter proposal in explaining all extant observations.

Horizon problem

The horizon problem results from the premise that information cannot travel faster than light. In a universe of finite age this sets a limit—the particle horizon—on the separation of any two regions of space that are in causal contact. The observed isotropy of the CMB is problematic in this regard: if the universe had been dominated by radiation or matter at all times up to the epoch of last scattering, the particle horizon at that time would correspond to about 2 degrees on the sky. There would then be no mechanism to cause wider regions to have the same temperature.

A resolution to this apparent inconsistency is offered by **inflationary theory** in which a homogeneous and isotropic scalar energy field dominates the universe at some very early period (before baryogenesis). During inflation, the universe undergoes exponential expansion, and the particle horizon expands much more rapidly than previously assumed, so that regions presently on opposite sides of the observable universe are well inside each other's particle horizon. The observed isotropy of the CMB then follows from the fact that this larger region was in causal contact before the beginning of inflation.**Heisenberg's uncertainty principle** predicts that during the inflationary phase there would be **quantum thermal fluctuations**, which would be magnified to cosmic scale. These fluctuations serve as the seeds of all current structure in the universe.[79]:207 Inflation predicts that the **primordial fluctuations** are nearly **scale invariant** and **Gaussian**, which has been accurately confirmed by measurements of the CMB.

If inflation occurred, exponential expansion would push large regions of space well beyond our observable horizon.

A related issue to the classic horizon problem arises because in most standard cosmological inflation models, inflation ceases well before **electroweak symmetry breaking** occurs, so inflation should not be able to prevent large-scale discontinuities in the **electroweak vacuum** since distant parts of the observable universe were causally separate when the **electroweak epoch** ended.

Magnetic monopoles

The magnetic monopole objection was raised in the late 1970s. Grand unified theories predicted topological defects in space that would manifest as magnetic monopoles. These objects would be produced efficiently in the hot early universe, resulting in a density much higher than is consistent with observations, given that no monopoles have been found. This problem is also resolved by cosmic inflation, which removes all point defects from the observable universe, in the same way that it drives the geometry to flatness.

Flatness problem

The overall geometry of the universe is determined by whether the Omega cosmological parameter is less than, equal to or greater

than 1. Shown from top to bottom are a closed universe with positive curvature, a hyperbolic universe with negative curvature and a flat universe with zero curvature.

The flatness problem (also known as the oldness problem) is an observational problem associated with a Friedmann–Lemaître–Robertson–Walker metric. The universe may have positive, negative, or zero spatial curvature depending on its total energy density. Curvature is negative if its density is less than the critical density, positive if greater, and zero at the critical density, in which case space is said to be *flat*. The problem is that any small departure from the critical density grows with time, and yet the universe today remains very close to flat. Given that a natural timescale for departure from flatness might be the Planck time, 10^{-43} seconds,[4] the fact that the universe has reached neither a heat death nor a Big Crunch after billions of years requires an explanation. For instance, even at the relatively late age of a few minutes (the time of nucleosynthesis), the universe density must have been within one part in 10^{14} of its critical value, or it would not exist as it does today.

Ultimate fate of the universe

Main article: Ultimate fate of the universe

Before observations of dark energy, cosmologists considered two scenarios for the future of the universe. If the mass **density** of the universe were greater than the **critical density**, then the universe would reach a maximum size and then begin to collapse. It would become denser and hotter again, ending with a state similar to that in which it started—a Big Crunch. Alternatively, if the density in the universe were equal to or below the critical density, the expansion would slow down but never stop. Star formation would cease with the consumption of interstellar gas in each galaxy; stars would burn out leaving **white dwarfs**, **neutron stars**, and **black holes**. Very gradually, collisions between these would result in mass accumulating into larger and larger black holes. The average temperature of the universe would asymptotically approach **absolute zero**—a **Big Freeze**. Moreover, if the proton were **unstable**, then baryonic matter would disappear, leaving only radiation and black holes. Eventually, black holes would evaporate by emitting **Hawking radiation**. The **entropy** of the universe would increase to the point where no organized form of energy could be extracted from it, a scenario known as heat death.

Modern observations of accelerating expansion imply that more and more of the currently visible universe will pass beyond our **event horizon** and out of contact with us. The eventual result is not known. The ΛCDM model of the universe contains dark energy in the form of a **cosmological constant**. This theory suggests that only gravitationally bound systems, such as galaxies, will remain together, and they too will be subject to heat death as the universe expands and cools. Other explanations of dark energy, called **phantom energy** theories, suggest that ultimately galaxy clusters, stars, planets, atoms, nuclei, and matter itself will be torn apart by the ever-increasing expansion in a so-called **Big Rip**.

Speculations

Main article: Cosmogony

While the Big Bang model is well established in cosmology, it is likely to be refined. The Big Bang theory, built upon the equations of classical general relativity, indicates a **singularity** at the origin of cosmic time; this **infinite energy density** is regarded as impossible in **physics**. Still, it is known that the equations are not applicable before the time when the universe cooled down to the **Planck temperature**, and this conclusion depends on various assumptions, of which some could never be experimentally verified. *(Also see* **Planck epoch.***)*

One proposed refinement to avoid this would-be singularity is to develop a correct treatment of **quantum gravity**.

It is not known what could have preceded the hot dense state of the early universe or how and why it originated, though speculation abounds in the field of **cosmogony**.

Some proposals, each of which entails untested hypotheses, are:

Models including the **Hartle–Hawking no-boundary condition**, in which the whole of space-time is finite; the Big Bang does represent the limit of time but without any singularity.

Big Bang lattice model, states that the universe at the moment of the Big Bang consists of an infinite lattice of **fermions**, which is smeared over the **fundamental domain** so it has rotational, translational and gauge symmetry. The symmetry is the largest symmetry possible and hence the lowest entropy of any state.

Brane cosmology models, in which inflation is due to the movement of branes in string theory; the pre-Big Bang model; the ekpyrotic model, in which the Big Bang is the result of a collision between branes; and the cyclic model, a variant of the ekpyrotic model in which collisions occur periodically. In the latter model the Big Bang was preceded by a Big Crunch and the universe cycles from one process to the other.

Eternal inflation, in which universal inflation ends locally here and there in a random fashion, each end-point leading to a *bubble universe*, expanding from its own big bang.

Proposals in the last two categories, see the Big Bang as an event in either a much larger and older universe or in a multiverse.

Religious and philosophical interpretations

Main article: Religious interpretations of the Big Bang theory

As a description of the origin of the universe, the Big Bang has significant bearing on religion and philosophy. As a result, it has become one of the liveliest areas in the discourse between **science and religion**. Some believe the Big Bang implies a creator, and some see its mention in their holy books, while others argue that Big Bang cosmology makes the notion of a creator superfluous.

See also
Cosmology portal
Physics portal
Big Crunch
Cosmic Calendar
Edgar Allan Poe's Big Bang speculation
Shape of the universe

Notes

1. There is no consensus about how long the Big Bang phase lasted. For some writers this denotes only the initial singularity, for others the whole history of the universe. Usually, at least the first few minutes (during which helium is synthesized) are said to occur "during the Big Bang."
2. Detailed information of and references for tests of general relativity are given in the article tests of general relativity.
3. It is commonly reported that Hoyle intended this to be pejorative. However, Hoyle later denied that, saying that it was just a striking image meant to emphasize the difference between the two theories for radio listeners.[54]
4. Strictly, dark energy in the form of a cosmological constant drives the universe towards a flat state; however, our universe remained close to flat for several billion years, before the dark energy density became significant.

CHAPTER 18

NO BIG BANG? QUANTUM EQUATION PREDICTS UNIVERSE HAS NO BEGINNING

Liza Zyga
Feb, 16

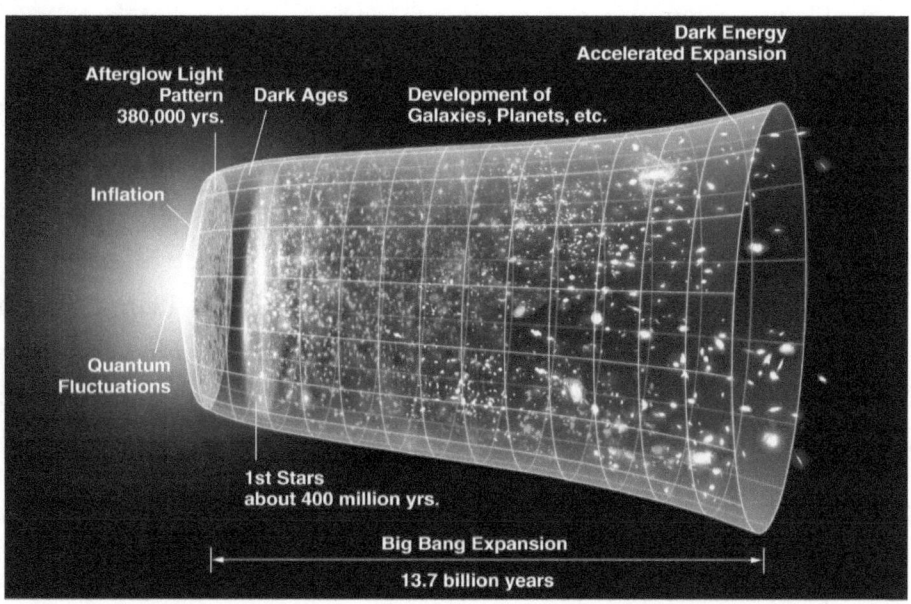

This is an artist's concept of the metric expansion of space, where space (including hypothetical non-observable portions of the universe) is represented at each time by the circular sections. Note on the left the dramatic expansion (not to scale) occurring in the inflammatory epoch, and at the center the expansion acceleration. The scheme is decorated with WMAP images on the left and with the representation of stars at the appropriate level of development. Credit: NASA

(Phys.org) — The universe may have existed forever, according to a new model that applies quantum correction terms to complement Einstein's theory of general relativity. The model may also account for dark matter and dark energy, resolving multiple problems at once.

The widely accepted age of the universe, as estimated by general relativity, is 13.8 billion years. In the beginning, everything in existence is thought to have occupied a single infinitely dense point, or singularity. Only after this point began to expand in a "Big Bang" did the universe officially begin.

Although the Big Bang singularity arises directly and unavoidably from the mathematics of general relativity, some scientists see it as problematic because the math can explain only what happened immediately after—not at or before—the singularity.

"The Big Bang singularity is the most serious problem of general relativity because the laws of physics appear to break down there," Ahmed Farag Ali at Benha University and the Zewail City of Science and Technology, both in Egypt, told *Phys.org*.

Ali and coauthor Saurya Das at the University of Lethbridge in Alberta, Canada, have shown in a paper published in *Physics Letters B* that the Big Bang singularity can be resolved by their new model in which the universe has no beginning and no end.

Old ideas revisited

The physicists emphasize that their quantum correction terms are not applied *ad hoc* in an attempt to specifically eliminate the Big Bang singularity. Their work is based on ideas by the theoretical physicist David Bohm, who is also known for his contributions to the philosophy of physics. Starting in the 1950s, Bohm explored replacing classical geodesics (the shortest path between two points on a curved surface) with quantum trajectories.

In their paper, Ali and Das applied these Bohmian trajectories to an equation developed in the 1950s by physicist Amal Kumar Raychaudhuri at Presidency University in Kolkata, India. Raychaudhuri was also Das's teacher when he was an undergraduate student of that institution in the '90s.

Using the quantum-corrected Raychaudhuri equation, Ali and Das derived quantum-corrected Friedmann equations, which describe the expansion and evolution of universe (including the Big Bang) within the context of general relativity. Although it's not a true theory of quantum gravity, the model does contain elements from both quantum theory and general relativity. Ali and Das also expect their results to hold even if and when a full theory of quantum gravity is formulated.

VisualDOC MDO Software

MDO, DOE, RSA, Probabilistic Design Easily automate design processes Go to vrand.com

No singularities nor dark stuff

In addition to not predicting a Big Bang singularity, the new model does not predict a "big crunch" singularity, either. In general relativity, one possible fate of the universe is that it starts to shrink until it collapses in on itself in a big crunch and becomes an infinitely dense point once again.

Ali and Das explain in their paper that their model avoids singularities because of a key difference between classical geodesics and Bohmian trajectories. Classical geodesics eventually cross each other, and the points at which they converge are singularities. In contrast, Bohmian trajectories never cross each other, so singularities do not appear in the equations.

In cosmological terms, the scientists explain that the quantum corrections can be thought of as a cosmological constant term (without the need for dark energy) and a radiation term. These terms keep the universe at a finite size, and therefore give it an infinite age. The terms also make predictions that agree closely with current observations of the cosmological constant and density of the universe.

New gravity particle

In physical terms, the model describes the universe as being filled with a quantum fluid. The scientists propose that this fluid might be composed of gravitons—hypothetical massless particles that mediate the force of gravity. If they exist, gravitons are thought to play a key role in a theory of quantum gravity.

In a related paper, Das and another collaborator, Rajat Bhaduri of McMaster University, Canada, have lent further credence to this model. They show that gravitons can form a Bose-Einstein condensate (named after Einstein and another Indian physicist, Satyendranath Bose) at temperatures that were present in the universe at all epochs.

Motivated by the model's potential to resolve the Big Bang singularity and account for dark matter and dark energy, the physicists plan to analyze their model more rigorously in the future. Their future work includes redoing their study while taking into account small inhomogeneous and anisotropic perturbations, but they do not expect small perturbations to significantly affect the results.

"It is satisfying to note that such straightforward corrections can potentially resolve so many issues at once," Das said.

Explore further: Theorists apply loop quantum gravity theory to black hole. More information: Ahmed Farag Ali and Saurya Das. "Cosmology from quantum potential." *Physics Letters B*. Volume 741, 4 February 2015, Pages 276–279. DOI: **10.1016/j.physletb.2014.12.057.** Also at: **arXiv:1404.3093**[gr-qc].

Saurya Das and Rajat K. Bhaduri, "Dark matter and dark energy from Bose-Einstein condensate", preprint: **arXiv:1411.0753**[gr-qc]

Journal reference: Physics Lettes B

CHAPTER 19

TOP TEN MIND BENDING THEORIES – TOPTENZ.NET

www.toptenznet/top-10-mind-bending-theories that try to answer some of these questions as mind bending in their own right and could change your......s

The universe is a vast and mysterious place. There are so many mind-boggling parts of our solar system, the universe and beyond that humans often have a hard time wrapping their brains around it. These mysteries are profound and often are connected with the inner workings of the universe and the very existence of life. Needless to say, a lot of these theories that try to answer some of these questions are mind bending in their own right and could change your whole outlook on reality.

10. We Don't Know How Many Planets Are In the Solar System

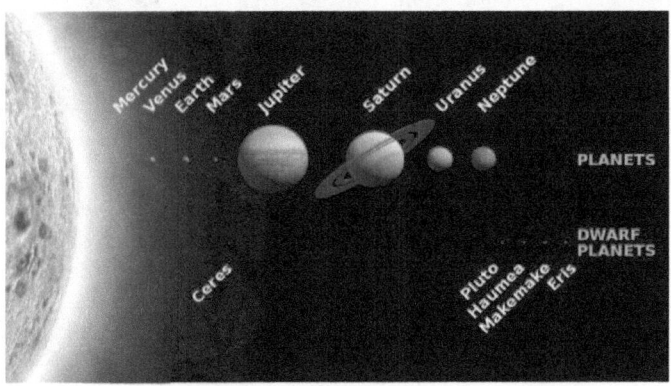

In our solar system, we have eight known planets and five known dwarf planets. We have known about the eight planets and Pluto for the past few centuries, but we still don't know what is beyond Pluto and still within our solar system, meaning there could quite possibly be more planets that we haven't discovered yet.

The first evidence of another planet within our solar system first arose after the discovery of two possible dwarf planets – 2012 VP113 and 90377 Sedna. Researchers noticed that something massive was affecting the orbit of these two possible dwarf planets. Astrophysicists believe that beyond the known planets there is a planet 10 times larger than Earth that is affecting their orbit. Beyond that, they believe that there are one or possibly more planets that are more massive than Earth, meaning that it is highly possible that our solar system has at least 10 planets.

These planets would be at least 200 astronomical units away from the Earth. One astronomical unit is the distance of the Sun to the Earth, which is about 93 million miles (150 million kilometers). Due to the great distance, we simply are unable to detect these other possible planets with current instruments.

9. Biological SETI

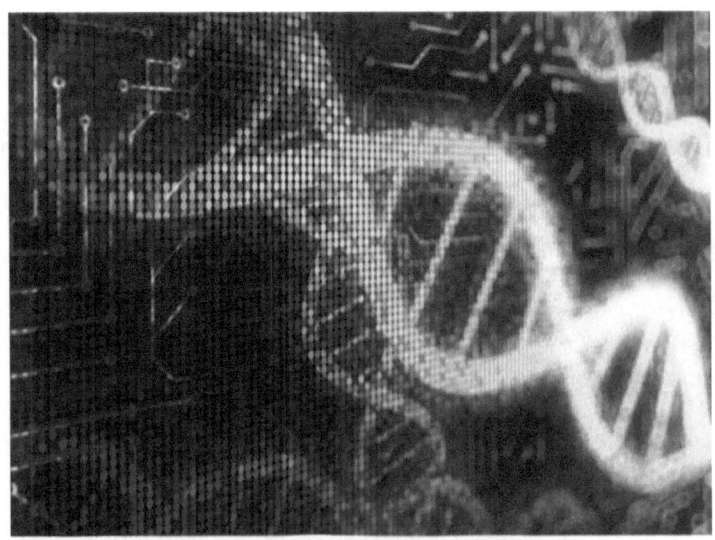

The biological SETI theory is an interesting pairing of two profound questions about human life. How did humans come into existence and are we alone in the universe? After all, life developing to this point is quite miraculous. Conditions had to be just right for present day humans to exist. Also, the idea that there are other beings living out in space has fascinated people for centuries.

However, what if we weren't meant to look to the skies for proof of alien life? What if it was written into our own DNA? That is what Cladimir Shcherbak of al-Farabi Kazakh National University of Kazakhstan, and Maxim Makukov of the Fesenkov Astrophysical Institute hypothesizes. Their theory is that DNA is one of the most durable constructs in the universe. If intelligent life were sending a message, it would be more effective to code it in our DNA, rather than send something like radio transmissions. Essentially, they are arguing that if there are cells in the human genome that cannot be explained by Darwinian evolution, that it is possible those cells are a signature, or a designer tag. They also point out how amazingly logical the human genome is. They believe that something as straightforward and logical, probably came from an advance being, somewhere outside of the solar system. If their theory is correct, then it could possibly answer both questions poised in the opening of this entry; there could be other life in the universe, and they could be the reason we exist.

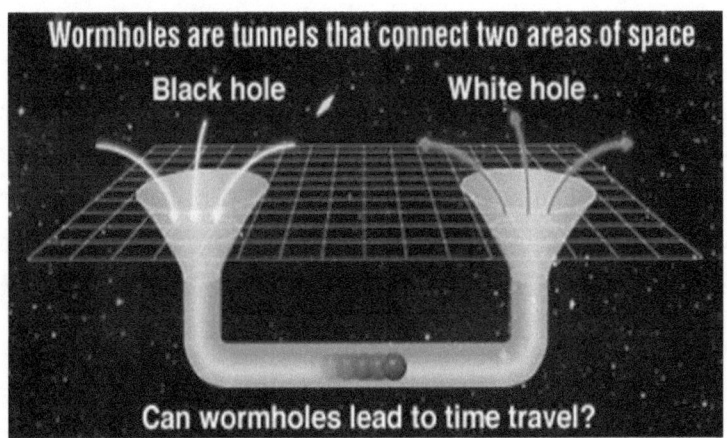

Most people have heard of black holes, but what about white holes? In the field of physics, white holes are a bit mythical; one has never been found, but it is possible for them to exist. The theory is that they are very much like a black hole, except for one main difference; you can't leave a black hole once you enter and you can only leave a white hole, you can't enter them. This means that black holes are entrances only, while white holes are exits only. Due to this binary, some physicists believe that black holes and white holes are actually the ends of wormholes. Meaning that someone, or something, could enter in the black hole and end up in some other part of space or in a completely different time once they come out the white hole.

However, this is all speculative because, as mentioned before, a white hole has never been found. In fact, some physicists believe that it is unlikely that they exist. They believe that if you sent something into a black hole, it would just get stuck there and eventually the black hole will get jammed up. Nevertheless, according to Einstein's theory of relativity, white holes are possible.

https://youtu.be/ZuvK-od647

7. Quantum Entalgement

Quantum Entanglement is a strange, physical phenomenon where tiny particles, like electrons, that were formerly entangled, interact with each other even after they are separated. This is seen when physicists measure a particle, because when it is measured it changes the property of the particle. The interesting thing is that if you change one particle, it also changes the one it was formerly entangled with. This is true, even if the two particles are a galaxy apart. It was such a problem for Einstein, that he called it "spooky actions at a distance," because it happened instantly, which is faster than the speed of light and went against his theory of relativity.

The leading theory is that, somehow, the two particles do manage to communicate with each other. The question then arises; can we communicate at faster than light speed using quantum entanglement? Unfortunately, that will not happen because quantum entanglements are random and we will not be able to send a message.

Understandably, this is a complicated subject, which even troubled Albert Einstein. If you wish to learn more about this topic, please watch the video posted above.

6. Baby Universes are Born in Black Holes 1

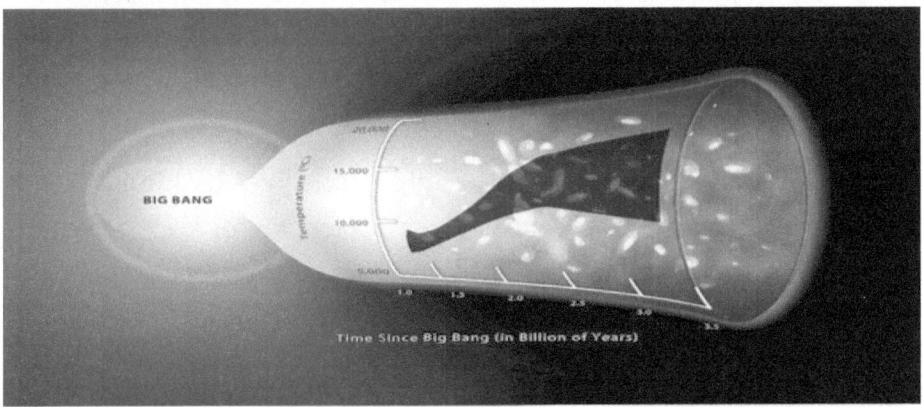

One of the most mysterious aspects of the known universe are black holes. There are a number of theories about what they are and what they do, but no one is exactly sure. One theory is that black holes are actually mothers that give birth to "baby universes." What's interesting is if the theory is true, then our universe is inside a black hole, which is in a black hole, which is in a black hole and so on.

While it is still a theory, it does help explain one mystery about the universe. Knowing the universe's rate of expansion, the universe is actually bigger than it should be. Physicists believe that in the fraction of a second after the Big Bang, it picked up power before tapering off. This fraction of a second is a period known as "inflation." No one is sure what caused the inflation, but it's quite possible that particles within a black hole could have caused the inflation. Without going into too much details about black holes, basically there are spinning half particles in them. When these half particles interact with each other, it creates a repulsive force called torsion. The torsion would explain the inflation because it would have given the Big Bang a little bit of extra thrust.

Black holes being mothers also helps explain why time works the way it does. In its simplest terms, our universe inherited the arrow of time from the mother black hole.

5. Time-Space is a Slippery Fluid

Spacetime is the concept where time and space are connected. For example, for humans on Earth experience space in three dimensions and the fourth dimension is time. The combination of these four dimensions would be considered spacetime.

Have you ever thought what spacetime feels like? It's kind of a silly question because space and time don't have a feel, right? Well, according to physicists, they believe that spacetime is actually a slippery substance referred to as superfluid. A superfluid is a type of fluid that has almost no friction or viscosity. One substance that is a superfluid is liquid helium when it is cooled to two degrees just above the coldest possible temperature, absolute zero. Essentially, a superfluid has so little friction that it makes water seem like molasses.

If spacetime were actually fluid, it would help align two branches of physic's, quantum physics and Albert Einstein's general theory. Neither of them can properly describe what happens to particles when spacetime undergoes a drastic change, like at the birth of a black hole.

4. The Ekypyrotic Scenario

The prevailing theory of how the universe came into existence was the Big Bang. In that first second after the Big Bang, most of the universe was created and it is still expanding. A mind-blowing question is, what sparked the Big Bang? According to the Ekypyrotic universe theory by Paul Steinhardt, a physics professor at Princeton it was caused by the collusion of two three-dimensional worlds (branes) in a space with a fourth dimension. According to the theory, two flat three-dimensional worlds (like strings) collided with each other and stuck together. This collision caused a kinetic energy to create a flat big bang universe, which is the universe in which we live. However, this goes against the mainstream belief that the Big Bang erupted out of singularity. If the theory of an Ekypyrotic universe is correct, it raises interesting questions about what is the outside of universe.

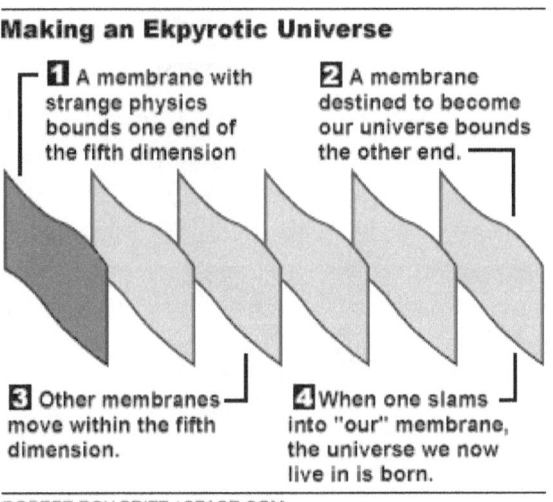

3. The Universe is Lopsided

In a fraction of a second after the Big Bang, there was a large burst, which gave the Big Bang a bit more thrust. This thrust is called inflation. It was believed that this inflation made the afterglow of the Big Bang, called the cosmic microwave background, pretty similar throughout the universe. However, data from NASA's Wilkinson Microwave Anisotropy Probe and the European Space Agency's Planck satellite show that this may not be true. In fact, according to the data, one side of the known universe is warmer than the other. The result is that the universe is lopsided.

Two theories as to why the universe is asymmetrical are that there is an energy field that is warping our universe. Another theory is that it is warped because it is bruised from bumping into another an universe or universes. Researchers said that if the universe is lopsided it probably will not really affect our understanding of the fundamental laws of physics, but again goes back to the question of what is outside our universe

2. Our Universe May Collide With Another One

The multiverse theory is a fairly well known, and mind-boggling, theory. Essentially, there are other universes that are parallel to our own. One of the best analogies to explain it is like apartment floors; they are all connected, but separate. The problem is what happens if one of those universes doesn't stay parallel and collides with our own? If that were to happen, we probably wouldn't be able to tell because it would hit us at the speed of light. However, if we were able to slow it down, it would be like a giant mirror coming from the sky. That would be the last thing we see, because after the collision, we would all die.

The good news is that is a worse case scenario and there is only a small probability that another universe will collide with us. Some physicists also believe that it is possible a collision with another universe has already happened.

1. The Fermi Paradox

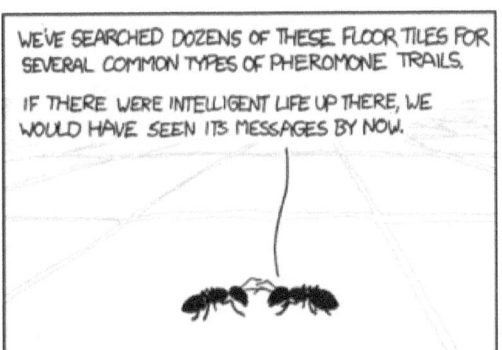

The Fermi paradox in its simplest terms is "where is everybody?" Why have we not come across some trace of an advanced civilization? There are four basic pillars to the argument. The first is that our solar system isn't that special, there are billions of other stars in the galaxy. If you compare the age of the sun to other stars, our sun is fairly young, by billions of years. Since our sun is just a normal star and if the Earth is just a normal planet, it is quite possible that there could be other Earth-like planets. Out of all those planets, even if a few developed interstellar travel at a pace similar to humans, then it would take a few tens of millions of years for just one of them to colonize the entire universe. Yet, despite the seemingly high probability that there should be some form of life at some point in the 13.8 billion year history of the universe, we have never found a single definitive artifact or evidence that there is anyone else in this universe except for us.

This leads to a few brain-bending theories, such as the zoo hypothesis, which is the notion that alien life is purposely staying away and could be observing us without our knowledge. Or that life is simply a computer simulation and the program is simply written where we're the only ones in the universe.

Robert Grimminck is a Canadian crime-fiction writer. You can follow him on Facebook, onTwitter, or visit his website

CHAPTER 20

A UNIVERSE FROM NOTHING?

by Jake Hebert, Ph.D.*

Explaining the origin of the universe is an enormous challenge for those seeking to deny their Creator: How could a universe come from nothing? The challenge is so great that some have argued that the universe simply did not even have a beginning, but has existed eternally. However, because most professing atheists have accepted the big bang model of the universe, they have accepted the premise that our universe did indeed have a beginning. Hence, they have a need to explain that beginning.

Theoretical physicist Lawrence Krauss presented in a recent book his claim that the laws of physics could have created the universe from nothing.[120] Likewise, other physicists offer similar arguments.

They appeal to the well-known phenomena of "virtual particle" creation and annihilation. The spontaneous (but short-lived) appearance of subatomic particles from a vacuum is called a *quantum fluctuation*. These subatomic particles appear and then disappear over such short time intervals that they cannot be directly observed. However, the effects of these virtual particles *can* be detected; they are, for instance,

---

* Dr. Hebert is Research Associate at the Institute for Creation Research and received his Ph.D. in Physics from the University of Texas at Dallas.

Cite this article: Hebert, J. 2012. A Universe from Nothing? *Acts & Facts.* 41 (7): 11-13.

[120] Krauss, L. 2012. *A Universe from Nothing.* New York: Free Press.

responsible for a very subtle effect on the spectrum of the hydrogen atom called the "Lamb shift." The short lifetimes of these virtual particles are governed by the Heisenberg Uncertainty Principle (HUP), which says that a short-lived state cannot have a well-defined energy.

The HUP places a limit on the time that a quantum fluctuation can persist. The greater the energy of the fluctuation, the shorter the time that it may last. It is for this reason that virtual particles appear and then disappear after *very* short intervals.

Krauss and other evolutionary physicists argue that the universe itself is the result of such a quantum fluctuation. However, the HUP itself presents an apparent difficulty for this claim. One would intuitively expect the energy content of the entire universe to be enormous. Hence, even if one were to argue that the universe *did* "pop" into existence via a quantum fluctuation, the energy content of the universe would be so large that the corresponding time would be vanishingly small, and the newly born universe would then immediately vanish. It is, therefore, difficult to see how our enormous universe could have resulted from such a fluctuation.

Evolutionary physicists argue, however, that if the total energy content of the universe were *exactly zero*, then a universe resulting from such a fluctuation could persist indefinitely without violating the HUP. This is admittedly a clever argument. Have the "new atheists" found a genuinely convincing way to explain our universe's existence apart from God?

Not really. The argument hinges on the claim that the total energy of the universe is exactly zero, and this claim is based squarely on Big Bang assumptions. Stephen Hawking writes:

The idea of inflation could also explain why there is so much matter in the universe....The answer is that, in quantum theory, particles can be created out of energy in the form of particle/antiparticle pairs. But that just raises the question of where the energy came from. The answer is that the total energy of the universe is exactly zero.[121]

---

[121] Hawking, S. 1996. *A Brief History of Time*. New York: Bantam Paperbacks, 133.

A Universe from Nothing?

Despite Hawking's blithe assertion, no human being can possibly know the precise energy content of the entire universe. In order to verify the claim that the total energy content of the universe is exactly zero, one would have to account for *all* the forms of energy in the universe (gravitational potential energy, the relativistic energies of all particles, etc.), add them together, and then verify that the sum really *is* exactly zero. Despite Hawking's intelligence and credentials, he is hardly omniscient.

So the claim of a "zero energy" universe is based, not on direct measurements, but upon an *interpretation* of the data through the filter of the Big Bang model. As hinted in the above quote, the claim comes from *inflation theory*, which states that the universe underwent a short, accelerated period of expansion shortly after the Big Bang. But "inflation" is an *ad hoc* idea that was attached to the original Big Bang model in order to solve a number of serious (and even fatal) difficulties.[122] Hawking, Krauss, and others are making the claim of a zero energy universe because it is an expected consequence of inflation theory. However, for someone who does not have an *a priori* commitment to the Big Bang (and inflation theory), it is not at all clear that the universe's total energy would be exactly zero. In fact, it seems extremely unlikely.

Moreover, when virtual particles momentarily appear within a vacuum, they are appearing in a space that already exists. Because space itself is part of our universe, the spontaneous creation of a universe requires space itself to somehow pop into existence.

In his recent book, Krauss spends very little time addressing this key point. Most of the book consists of a defense of the Big Bang, anecdotal stories, and criticisms of creationists. It is only near the end of the book that he actually seriously addresses this key issue (how space itself could be created from nothing), but he spends very little time on it, despite the fact that the book is over 200 pages long.[123] He argues that *quantum*

---

[122] Williams, A. and J. Hartnett. 2005. *Dismantling the Big Bang*. Green Forest, AR: Master Books, 121-125.

[123] Krauss, *A Universe From Nothing*, 161-170.

gravity (a theory that merges quantum mechanics and general relativity) could allow space itself to pop into existence. One obvious problem with this claim is that a workable theory of quantum gravity does not yet exist.

Moreover, the general claim that the laws of physics could have created our universe suffers from a number of serious logical difficulties. Our understanding of the laws of physics is based on *observation*. For instance, our knowledge of the laws of conservation of momentum and energy come from observations made from literally thousands of experiments. No one has ever observed a universe "popping" into existence. This means that any laws of physics that would allow (even in principle) a universe to pop into existence are completely outside our experience. The laws of physics, as we know them, simply are not applicable here. Rather, the spontaneous creation of a universe would require higher "meta" or "hyper" laws of physics that might or might not be anything like the laws of physics that we know.

But this raises another problem. Since such hypothetical meta or hyper laws of physics are completely outside our experience, why do atheistic physicists naively assume that rules like the HUP would even apply when describing the universe's creation? They freely speculate about other (unobservable) universes in an alleged "multiverse" that can have laws of physics radically different from our own. Since the HUP is known to be valid only *within* or *inside* our universe, it is not at all clear why they would assume that the HUP would even apply when discussing our universe's creation. Perhaps the HUP is indeed part of these hyper laws of physics, but one could just as easily argue that it is not. One can engage in all kinds of speculation here, but such speculation is not *science*.

Moreover, even if these supposed higher laws of physics actually existed, in order for them to create the universe, they must have an existence *apart* from the universe. But this presents a dilemma for the atheist who says that the cosmos is all that exists. Before his death, Carl Sagan acknowledged in correspondence with ICR scientist Larry

Vardiman that he recognized this problem for his worldview: His view of origins required the laws of physics to *create* the cosmos, but because he did not acknowledge his Creator, he could not explain the origin of the laws themselves.[124] The existence of physical laws external to the cosmos itself was an obvious violation of his well-known axiom "The Cosmos is all that is or ever was or ever will be."[125]

Of course, the atheist could try to dodge this difficulty by resorting to the claim that the cosmos simply had no beginning and is eternal.

But even this avoidance leaves unresolved difficulties. For instance, some are claiming that the cosmos as a whole—the so-called "multiverse"—is eternal, but that it contains infinitely many *individual* universes (a consequence of modern inflation theory). According to this view, it is only *our* particular universe that began 13.7 billion years ago. The existence of other alleged (but unobservable) universes supposedly explains our seemingly improbable existence—because the multiverse contains infinitely many universes, the laws of physics and chemistry in at least some of these universes would have properties necessary for life. Thus, our existence is supposedly explained because we just happen to live in such a universe.

A glaring fallacy exposes this argument: While the laws of physics and chemistry in our universe do indeed allow life to *exist*, they do *not* allow life to *evolve*. The laws of physics and chemistry simply are *not* favorable to the evolution of life.

For decades, creationists have pointed out the insurmountable difficulties with "chemical evolution" scenarios.[126,127,128] these difficulties don't vanish simply because someone claims that other (unobservable)

---

[124] Vardiman, L. 2012. Did the "God Particle" Create Matter? *Acts & Facts.* 41 (3): 12-14.

[125] Sagan, C. 1985. *Cosmos.* New York: Ballantine Books, 1.

[126] McCombs, C. 2004. Evolution Hopes You Don't Know Chemistry: The Problem of Control. *Acts & Facts.* 33 (8).

[127] McCombs, C. 2004. Evolution Hopes You Don't Know Chemistry: The Problem with Chirality. *Acts & Facts.* 33 (5).

[128] McCombs, C. 2009. Chemistry by Chance: A Formula for Non-Life. *Acts & Facts.* 38 (2): 30.

universes exist. Even if the laws of physics and chemistry in *every single* one of these other supposed universes did allow for life to evolve, those laws from another universe could not explain the existence of life in *this* universe. This should have occurred to the atheists—but their argument demonstrates "vain imaginations" and "foolish, darkened hearts" (Romans 1:21-23).

Despite the impressive academic credentials of those promoting the "universe from nothing" idea, the scenario is utterly unreasonable, and no Bible-believing Christian should be intimidated by these "vain imaginations."

Physics Q&A

A Universe from Nothing? (**file:///article/6901**)

Has Einstein's Limit on the Speed of Light Been Broken? (**file:///article/69**04)

Have Scientists Found the 'God' Particle? (**file:///article/6914**)

Proton Problems: Faith in Theories or Reality? (**file:///article/8018**)

The Higgs Boson: A Blow to Christianity? (**file:///article/6925**)

Where Is the Universe Headed—to Order or Chaos? (**file:///article/5823**)

Related Articles

A Universe from Nothing? (file:///article/9687)

Proton Problems: Faith in Theories or Reality? (**file:///article/8018**)

The Higgs Boson: A Blow to Christianity? (**file:///article/6925**)

Physics, Not God, Explains the Universe? (**file:///article/6912**)

Have Scientists Found the 'God' Particle? (**file:///article/6914**)

CHAPTER 21
MULTIVERSE (RELIGION)

A multiverse is the concept of a plurality of universes. Some religious cosmologies propose that our universe is not the only one that exists.

Srimad Bhagwatham

विकारैः सहितो युक्तैविशेषादिभिरावृतः ।
आण्डकोशो बहिरयं पञ्चाशत्कोटिविस्ततः ॥

(XI:16:39)

देशोत्तराधिकैर्यत्र प्रविष्टः परमाणुवत् ।
लक्ष्यतेदन्तर्गताश्चान्ये कोटिशोह्याण्डराशयः ॥

(XI:16:40)

This Brahmanda (Cosmic shell), which consists of the eight special modifications and sixteen effects of Prakriti, has a dimension of fifty crores of Yojanas (9 miles make one yojana) within, and is surrounded without by the layers of the five elements, the layer of each element being ten times the dimension of the internal dimension of the preceding Shell. All this is only like an atomic particle in Sri Hari. In Him there are countless other Shells like this, all of whch together even is like a few atomic particles for Him.

क्षित्यादिभिरेष किलावृतः सप्तभिर्दशगुणोत्तरैराण्डकोशः ।
यत्र पतत्यणुकल्पः सहाण्डकोटिकोटिभिस्तदनन्तः ॥

(VI:16:37)

This cosmos is encased in seven sheaths of earth, water, fire, air, sky, I-sense (Ahankara), and the great element (Mahattattva), each succeeding sheath being ten times vaster than the earlier one. Amidst countless such world systems (Brahmandas), this cosmic shell or world system of ours is whirling about like an atom in the infinitude of Thyself – the Ananta or the Endless One.

वयं न तात प्रभवाम भूम्नि यस्मन् परेऽन्येऽप्यजीवकोशाः ।
भवन्ति काले न भवन्ति हीदृशाः सहस्त्रशो यत्र वयं भ्रमामः ॥

(IX:4:56)

Sri Rudra said: He by whose will innumerable universes (Brahmandas) like the one in which we are involved, emerge and get dissolved continuously, – against Him, we are not capable of doing anything.

क्वाहं तमोमहदहंखचराग्निवार्भूसंवेष्टिताण्डघटसप्तवितस्तिकायः ।
क्वेदग्विधाविगणिताण्डपराणुचर्यावाताध्वरोमविवरस्य च ते महित्वम् ॥

(X:14:11)

What am I, a being with a limited body of seven spans – this Brahmanda (Cosmic Shell) of seven layers of Prakriti's evolutes, the Mahat, Ahankara and the five elements – what am I of such description before Thee, whose every pore is like a window through which countless such Brahmandas are passing like particles of dust.

य्दुपतय ए ते न ययुरन्तमनन्ततया त्वमपि यदन्तराण्डनिचया नन सावरणाः ।
ख इव रजांसि वान्ति वयसा सह यच्छतयस्त्वयि हि फलन्त्यतन्निरसनेन
भवन्निधनाः ॥

(X:87:41)

Even leading Divinities like Brahma do not find the limits of Thee; for Thou art the unlimited, and the unlimited by its nature cannot be fully comprehended. For the same reason even Thou dost not know Thy limits. Within that infinite and incomprehensible being of Thine, countless Brahmandas (universes), each with its seven expansive external coverings, whirl about together under the propulsion of Time like clusters of dust in the air. So the words of the Veda, unable to describe Thee positively, arrive at Thee only as the final residue left after negating all conceivable entities.

सख्यानं परमाणूनां कालेन क्रियते मया।
न तथा मे विभूतीनां सृजतोऽण्डानि कोटिशः ॥३९॥

It may be possible for Me to count the total number of the primary atoms in the course of a vast period of time. But even I cannot exhaustively count and estimate My powers and glories; for out of them crores and crores of world systems are always originating, thus setting no limits to them.

The concept of multiple universes is mentioned many times in Hindu Puranic literature, such as in the Bhagavata Purana:

The concept of parallel universes appears in the Brahma Vaivarta Purana:

*And who will search through the wide infinities of space to count the universes **side by side**, each containing its Brahma, its Vishnu, its Shiva? Who can count the Indras in them all – those Indras **side by side**, who reign at once in all the innumerable worlds; those others who passed away before them; or even the Indras who succeed each other in any given line, ascending to godly kingship, one by one, and, one by one, passing away?* (Brahma Vaivarta Purana)

CHAPTER 22

IS THE UNIVERSE CONSCIOUS?

Steve Stewart-Williams

This post is excerpted, with changes, from the book *Darwin, God and the Meaning of Life* by Steve Stewart-Williams – available now from Amazon.com (http://www.amazon.com/gp/producy/0521762782)File:///E.%Is Universe Concious_Psychology.Today.htm

Take out a piece of paper and draw a vertical line right down the middle. On one side of the page, make a list everything in the world that you're sure is conscious; on the other, make a list everything that you're sure is not. Unless you're one of those people who pretends that consciousness doesn't exist, it's a safe bet that you'll have human beings on the conscious side; probably you'll have at least some other animals there to keep them company too. Unless you're a New Ager or mentally unbalanced, it's a safe bet that you'll have things like rocks and potato peelers on the other side, the unconscious side.

One thing you probably won't have on your list of conscious things is: the universe. To most people (myself included), the idea that the universe might be viewed as a conscious entity is outlandish in the extreme – the kind of view that most scientifically-minded people (myself included) would be unwilling to admit to, at least around their scientifically-minded friends and colleagues. But as outlandish as it might sound, it actually turns out to be a straightforward implication of evolutionary theory.

Is the Universe Conscious?

This is the case, at least, if we assume that not only the body but also the mind is a product of evolution. And this is a view that really everyone should hold today. We know that the mind is dependent on the activity of the brain, and we know that the brain is a product of evolution; thus, we know that the mind is a product of evolution. Furthermore, if you look at some of the core components of the human mind, they have clear links to survival and reproductive success: fear motivates the avoidance of danger; sexual desire motivates behaviours that lead to the production of babies; etc. The case that the mind is a product of evolution is strong.

As soon as we recognize that the mind is a product of evolutionary processes, our view of the mind and its place in nature is radically altered. A recurring idea in the belief systems of the world is the notion that human beings are composed of two separate and separable parts: a physical body and an immaterial mind or soul. Not everyone has held this view, but it's not (as commonly claimed) unique to the Western tradition. Many people have believed that the mind stands outside nature; it is a part of us that transcends the material world and our biology.

Evolutionary theory completely overthrows this view. From an evolutionary perspective, it is impossible to maintain that the mind stands outside nature. Instead, it is a tiny fragment of nature, valued only by those tiny fragments of nature that possess it. Mind is not something separate from matter; it is a process embodied in matter.

Now here's the point: When we fully digest that the mind is the activity of an evolved brain, it radically transforms our view of the mind's place in the universe – and our view of the universe itself. The physical universe ceases to be an unconscious object, observed and explored by conscious minds which somehow stand above or outside it. Conscious minds are *part* of the physical universe, as much as rocks and potato peelers. Our consciousness is not simply consciousness *of* the universe; our consciousness is a part of the universe, and thus the universe itself is partially conscious. When you contemplate the universe, part of the universe becomes conscious of itself.

Similarly, our knowledge of the universe is not something separate from the universe; it is a part of the universe. Thus, for humans to know the universe is for the universe to know itself. As Carl Sagan put it, 'humans are the stuff of the cosmos examining itself.' And Darwin's theory of evolution explains how this could be so – how clumps of matter could come to be organized in such a way that they are able to contemplate themselves and the rest of the cosmos.

The history of the universe looks very different from this perspective. For billions and billions of years, the universe was here and no one knew about it. More to the point, for all that time, the universe itself had no idea that it existed. But then, around 13.82 billion years after the Big Bang, and almost four billion years since life first evolved, something strange began to happen: Tiny parts of the universe became conscious, and came to know something about themselves and the universe of which they are a part. Eventually, some of these tiny parts of the universe – the parts we call 'scientists' and 'scientifically-informed laypeople' – came to understand the Big Bang and the evolutionary process through which they had come to exist. After an eternity of unconsciousness, the universe now had some glimmering awareness that it existed and some understanding of where it had come from. This might sound like a strange thing for a universe to do, but perhaps it's not; perhaps many possible universes would become conscious of themselves given sufficient time.

So much for the history of the conscious universe; what about its destiny? There are many competing suggestions on this topic, some more optimistic than realistic. Pierre Teilhard de Chardin, a French Jesuit palaeontologist, suggested that the universe will continue to expand into greater and greater degrees of awareness, finally coalescing into an integrated, universal consciousness, which he dubbed the Omega

Point and identified with Christ. Modern cosmology indicates that such suggestions are more interesting than they are plausible. Although the universe is conscious of itself at present, the projected heat death of the universe makes it all but certain that the time will come when the lights go out and the universe slips back into unconsciousness.

For how long will it remain in its present semi-conscious state? The answer depends on how prolific the universe is at producing conscious life. If consciousness is widespread throughout the universe, then the odds are that at least some pockets of consciousness on some planets will survive for a reasonable length of time. For all we know, though, ours may be the only planet in the universe hosting mind and consciousness. If so, then our decisions and our conduct will determine whether the universe has a long future as a conscious entity or will soon lapse back into unconsciousness.

That said, one might wonder whether, in the grand scheme of things, it really matters. It may be pure anthropocentrism to assume that a universe with consciousness is better than one without. Conscious beings are often disgruntled and sometimes simply miserable, and maybe on balance an unconscious universe would be the more desirable. But although it's possible to entertain such thoughts in principle, it's hard in practice to duck the conclusion that it would be a terrible shame if the universe were not to remain conscious for as long as possible. Nonetheless, it may be the fate of the universe to spend an eternity in darkness, save one brief flash of self-awareness in the middle of nowhere.

CHAPTER 23

CONCLUSION

For accurate analysis of the present study of Creation of Universe – God or Big bang let us take help of Pramana (Evidence). No one was present to witness the Creation and hence **Pratyaksha**, which means perception arising from interaction of five senses and their wordly objects, does not apply to the study. **Anumana** which means inference can be applied when reaching a new conclusion from one or more observations. Anumana can also be applied to the scientific knowledge, observation and then coming to conclusion. Big Bang theory depends on Anumana. Again **Upamana** which means comparison and anology can again be applied to scientific observation and conclusion. **Arthapathi** means postulation, derivation from circumstances. Many scientific theories are postulated, immensely contributing to our present day knowledge and hence Arthapathi is valid in our present context. **Anupalabdi** means non perception, negation/cognitive proof. Many old scientific theories have become redundant on this basis.

Sabda (word) means relying on word, testimony of past or present. The Sanatana School which consider it epistemically valid suggest that a human being needs to know numerous facts, and with the limited time and energy available he can learn only a fraction of facts and truths directly. He must rely on others, his parents, teachers, ancestors etc. This means gaining knowledge and enrich each others live and also means proper knowledge is either spoken eg attending lectures or written through Sabda (words). In our present study we have to rely on Sabda (words) to a very great extent. Religious books being words of God/revealed, description of Creation in each of these books cannot be denied. Coming to Sanatana Dharma (Hinduism) there is no single book, reading which one becomes Hindu. There are six Sidhantas namely, Nyaya, Vaisesika, Sankhya, Purva Mimansa, Uttara Mimansa and Vedanta, 108 Upanishads, 108 Puranas and various other schools of thoughts like Advaita (non duality), Dvita (Duality), Vasistadvita etc. When it comes to Creation of Universe we have to completely rely on Books of Religion i.e., word of God be it Vedas, Bible or Quran.

Samkhya Sutra:

Sankhya or Samkhya philosophy is considered to be the most ancient school of thought. Samkhya as mentioned in previous paragraph forms the most important pillars of Indian philosophy (Siddhanta). Samkhya reduces everything to two entities – Prakriti (nature) and Purusa (God). Samkhya also describes three Gunas viz Sathvik (that which is pious and does good deeds), Rajas (Passion) and Tamas (darkness).

Samkhya is enumeristic philosophy which accepts 3 of six pramanas which include Pratyaksha (Perception), Anumana (inference) and Sabda (word). Samkhya philosophy is dualistic and regards universe as consistent of two realities; Purusa (Consciousness) and Prakriti (Matter). Jiva (living being is that state in which Purusha is bonded to Prakriti). This fusion of Purusha and Prakriti leads to emergence of Bhuddhi (interllect) and Ahankara (Ego).

The Samkhya system is based as Sat-Karya-vada or theory of Causation. According to this theory effect is pre-existent in the cause.

Ishvarakrishna in his Samkhyakarika gives five reasons, why effect has to be pre-exist in material cause –

1. What is not cannot be produced
2. The effect requires material cause

3. Not everything arises from everything
4. The cause produces only what corresponds to this potential and
5. The effect has nature of the cause.

Satkaryavada conceives possibilities based on historical documents. Advaita Vedanta who place their belief in Vivartavada which menas alteration, modification, change of form or state. Vivarta involves Vikara or modification but not apparent modification (of the real which does not change). Therefore, the world is vivartaa of the soul entity Brahman and merely an illusion. Brahma satyam Jagat mithya by Adi Shankaracharya (600 CE) means God is real and the world is unreal.

Classical Samkhya argues against the existence of God because they argue that unchanging God cannot be source of an everchanging world. The preceding pages give explations against Samkhya Sutra(Rebuttal).

Bhagwat Gita Ch 13 verse 13 says "Brahman is spirit beginningless and lies beyond cause and effect of material nature" thus arguing against Satkaravada of cause and effect.

Samkhya rejects the notion of God and its existence. Samkhya theorists argue that unchanging God cannot be source of everchanging world. To counter this Katha Upanishad 2.2.13 says "Brahman is the one unchanging ground of the entire phenomenal existence, which is super imposed upon it by avidya (nescience). The Lord is unchanging substratum of entire changing universe during its creation, preservation and dissolution without consciousness of atman, all beings would become inert."

Further Samkhya philosophy denying existence of God state that even if Karma (work, good or bad), is denied, God still cannot be enforcer of consequences. Because motive of an enforcer God would either is egoistic or altruistic.

To this Bh.G 5:15 says "Nor does supreme Lord assume anyone's sinful or pious activities." Bh. G 4:12 explains that men in this world desire success in fruitive activities and therefore they worship demigods.

Quickly, of course get results from fruitive work. Thus God is not enforcer.

Samkhya philosophy states that, there is no proof of existence of God. The sole purpose of Upanishads according to Adi Shankaracharya is to prove the reality of Brahman (God) and the phenomenality or unreality of the universe of names and forms.

Brihadaranyaka Upanishad describes Brahman in the following way,

1. Sat-chit-Anand

   Sat = Existence, chit = Consciousness, Anand = Bliss
2. Brahman is universal spirit, the ultimate reality, pure consciousness: the one existence, the Absolute "the unchanging reality amidst and beyond the universe", which "cannot be exactly defined.
3. Its secret name og zBrahman is Satyasya Satyam, 'The Truth of all truth (Bri Up2.1.20.)"I am the way and the truth and life. (John 14.6)
4. About instruction about Brahman: Neti, neti-not this, not this. By neti, neti one cannot prove existence of God. By neti, neti what remains in the end is you. For there is no other (Bri up 2.3.6 & cha up 3.14.1). "O Lord there is non like You, nor is there any God besides You" (Chronicles 17:20)

Samkhya Sutra Aph 19

"Says [But not without conjunction there of (i.e., of Nature) there is the conjunction of that [i.e., of pain] with that [viz, the soul] which is ever essential pure and free intelligence."

Bhagwat Gita says that a spirit embodied soul (Jivatma) which is bound by Karma (work) and spirit (Jivatma) get energised (get animated) by virtue of ever pervading consciousness (soul). The soul is witness, it does not take part in any action and reaping of fruits of action of spirit (Jivatma) and it is spirit which passes through the cycle of Birth-death-Rebirth.

Aphorism 79 states that "If [the world] is not unreal because there is no fact contrary [to this reality] and because it is not [false] result of depraved causes [leading to belief in what not to be believed.

Adi Shankaracharya says that Brahma Satyam jagat mithya which means that 'The Reality, Brahman which is free from all evil, which is pure consciousness, which is substratum of this illusory world, that I am." "Know, then that Prakriti is maya (illusion) and that God is real." (Svet. Up IV.10).

The sun, the moon and stars appear as round, flat and small. They appear to move slowly in sky. But in reality they are not attached to anything. They are not flat. They do not move from east to west. They remain in one place. Each planet is several times bigger than earth. The earth also appears flat. Therefore things are not as they appear to be. The visibles are illusionary.

"All entities are born in delusion"(Bha. G 7:27)

- Brahman projects the universe through the power of its maya" (Svet. Up. IV.9).

When thinker and thought are integrated through right meditation then there is ecstasy of the Real (Jiddu Krishnamurty)

Aph 92 [This objection to definition of Perception has no force], because it is not proved that there is a Lord.

This argument is like sending a man into a dark room in which there is a black cat and asking him "what do you see" and he says "I see nothing." Just because he cannot see a black cat in the dark room does not imply that there is no black cat and hence Aph 92 does not stand. God cannot be perceived by our five senses, organs which are limited in their capacities. Br Up IV iv.19 "By mind alone is Brahman to be realized. There exists in it no diversity whatsoever." "That the conclusion of the scriptures is that Brahman is formless"-Brahma Sutra.III.ii.14:Vyasa further emphasizes that undifferentiated nature of Brahman by stating that 'It is pure Consciousness.' It is characterized by negation (neti, neti) of all attributes. It is higher than the known and other than the unknown. It is devoid of all multiplicity and is completely other than phenomenal Universe (Br Su III.iv.18)

Aph 93 [and further] it is said that [the Lord] exists because [who ever exists must be either free or bound, and], of free and bound he can be neither the one nor other.

Bha Gita 13:14 "Brahman the spirit is beginningless and lies beyond cause and effect of this material world."

Aph 132, 135, 139, 141, 142, 143, 148

There is no dispute in vedatanta on the above aphorisms because Vedanta accepts Sat-Kara-Vada or theory of causation. According to this theory effect is pre-existant in cause.

Aph 149 From the several allotment of birth, a multiplicity of souls (is to be inferred).

Samkhya Sutra Siddhanta says that each individual hasa single soul. So multiple individuals must have multiple souls. To this Adi Shankaracharya says that Atman (Soul) is only one and unique. Atman alone (Ekaatma Vaadam). It is a false concept that several Atmans (Anekaatma vaadam(many souls)to which Samkhya Sutra professes. Adi Sankaracharya says that just as the same moon appears as several moons on its reflection on surface of water covered with bubbles, the One Atman appears as multiple Atma because of Maya (illusion).

"This individual soul is unbreakable and insoluble and can be neither burned nor dried. He is everlasting, present everywhere, unchangeable, immovable and eternally same."(Bha G 2:24)

The self (soul or Atman) is Brahman. That means there is no difference between Atman of humans and God(Bri Up). "I AM He who I Am"(Exo 3.14) "the Kingdom of Heaven is in your mist" (Luke 17.21) "Don't you know that you yourself are God and that God's spirit dwells your midst." Corianthians 3:16

Aph 157 already explained on commentary on Aph 149.

Book V of Sankhya Sutra

Aph 3 [If a Lord were governor, then] from intending his own benefit, his government (would be selfish) as in the case [with ordinary governors] in the world.

"There is no work prescribed for me within all the three planetary systems. Nor am I want anything, nor have I needed to obtain anything – and yet I am engaged in prescribed duties." (Bh. G: 3.22).

"If I did not perform prescribed duties, all these worlds would be put to ruination. I would be the cause of creating unwanted population and I would thereby destroy the peace of all living beings" (Bh. G 3:24). "My father is always working to this very day and I am too working" (John 5:17).

God who has been projected as Father in New Testament of Bible has been depicted as Compassionate God. Jesus always worked for the down trodded, performed many miracles by healing deaf, blind, leprosy, excessive bleeding in woman, raising dead, blind to see and many more. Jesus is the most Compassionate to have lived on this earth. No one in the history of the world has been as compassionate as Jesus.

Aph 4 [He must, then be] just like a world by lord, [and] otherwise [than you desire that we should conceive of him] meaning that God's desires must be just like a worldly lords.

"Intelligence, knowledge, freedom from doubt, and delusion, forgiveness and distress, birth and death, fearlessness, austerity, charity, fame and infamy – all these various qualities of living beings are created by Me alone"(Bh G 13:4-5.)"I am the source of all spirited and material world."(Bh. G 10:8)

The above has evidence that God is desireless and not liable to grief, when He is the source of all spirited and material world where is desire for him not satisfied. Verse Bh. G 13:4-5 above His forgiveness, fearlessness, non violence, equamity and satisfaction and charity created by Him then where is the question of grief in Him.

Conclusion

Aph 5 or [let the name of Lord be technical.

Meaning if, whilst there exists also world, there be a Lord, then let yours like ours, be merely a technical term for the soul which emerged at the commencement of Creation since, there cannot be eternal lordship, because of contradiction between mundaneness and the living unobstructed will.

"The individual soul is unbreakable and can neither burn nor dried. It is everlasting, present everywhere, unchangeable, immovable and eternally same." (Bh G 2:24).Atman exits even before creation of universe.

"Brahman (God) is beginningless and subordinate to Me, lies beyond the cause and effect of this material world"(Bh G 13:13)

Aph 6 This [position viz., that there is a Lord] cannot be established without assuming that he is affected by] Passion; because that is the determinate cause [of all energizing]. As per Samkhya Sutra passion is a mode in Prakriti and Prakriti is not Mahat (Great or God). Hence God cannot have passion or three modes named as Sathvik, Rajas and Tamas (passion) as described in Samkhya Sutra.

Aph 7 Moreover, were that [passion] conjoined with him, he could not be entirely free.

Bhag Gita ch 13 verse 32 "Those with vision of eternity can see that imperishable atman (which is also Brahman) is transcendental, eternal and beyond modes of nature."

This is to explain that modes of Prakriti such as Sathvik, Rajas and Tamas do not affect Him. The Atman which is Brahman is free from these three modes.

Aph 9 If these were from mere existence [of nature, not in association but simply in proximity], their lordship would belong to every one.

"The Supreme Truth (God) exists outside and inside of all living beings, the moving and non-moving (Bha G 13:16). He is subtle. He is beyond Power of material senses to see to know. Although He is far, far away, He is also near to all" (Bha G 16).

Quran also says "We (Allah) are closer to you than Jugular vein. Quran (50:16)

"He is the source of light in all luminous objects. He is beyond darkness and is unmanifest. He is knowledge. He is object of knowledge and He is goal of knowledge. He is situated in everyone's heart." (Bhag G14:18)

Aph 10 It is not established [that there is eternal Lord]; because there is no evidence of it.

That Brahman illuminates. It is Light of all lights that is the language. 'Natara' Süryo bhāti 'there the sun does not shine.' The sun illuminates the entire solar system but not Brahman. It is Brahman's power that makes the sun illuminate. 'na candratarakam nemā vidyuto bhānti, Kuto ayam agnit,' neither do moon and stars shine there, nor do flashes of lightening, what to speak of fire? They are nothing. They have no power at all to illuminate Brahman.

'Tam eva Bhāntam anubhati sarvam.' The Brahman shines. Tasya bhāsā sarvamidam vibhāti, "By His light the whole world is lighted. The sun, moon, the stars and lightening are lighted by the infinite Brahman of the native of pure consciousness" (Katha Upanishad 2.2.15).

Katha Upanishad says: "Sā Kāstha sā parā gati." That is the Supreme end, supreme destination. There is nothing higher than that.

Mandukya Upanishad says – All is Om (AUM) Om has three syllable A, U, M. Om can be divided into four quarters. A is associated with waking state, U is associated with dreaming state in sleep, M is associated with deep dreamless sleep. When all the three become one then there is Turiya state. In Turiya state one experiences infinite bliss. That Bliss is Brahman.

"Neti neti i.e., not this and not that. If you go on saying that God cannot be this and God cannot be this to infinitismal point (that is

Conclusion

everything that exists in this universe which is infinite) that is Brahman." (Bri. Up 4.4.22)

Brahman is not an object it is Adrisya (imperceptable), beyond the reach of eyes. It is not another. It is all-full, infinite, self-existent, self-delight, self-knowledge and self-bliss. It is Svarupa (Ones own shape), essence. It is the essence of knower. It is Drashta (seer), Turiya (Transcedendent) and Sakshi (Silent Witness).

Aph 11 There is no inferential proof [of there being a Lord]; because there is [here] no [case of invariable] association between sign and that which it might be betoken(be a sign of).

Brahman cannot be known by inference. It has to be experienced. You cannot prove Brahman by evidence (betoken).

Vedas depict Brahman as the Ultimate Reality, the Absolute, theBrahman is indescribable, incorporeal, omniscient, original, first, eternal, both transcendent and immenent, absolute infinite existence and ultimate principle who is without beginning, without end, who is hidden in all and who is cause, source of all material and effect of all creation known, unknown and yet to happen in the entire universe. Brahman is Kshetragna i.e., witness so where is the need to prove Brahman by evidence. He is knower & known does not require evidence.

Rig Veda:

Introduction: Of the four vedas Rig Veda (RV) is considered to be the oldest. The Sanskrit language is different from those of other 3 vedas viz., Sama, yajur and attarva. The Rig Vedic Sanskrit is different from Maha Kavi Kalidasa's Sanskrit. Many scholars were of the opinion that Rig Veda Sanskrit being different is also difficult. More than 10 western Indologists translated RV in English. The main two are MaxMuller and Rayon T. H. Griffith and 10 others. These translations by western Indologists did not have concept of Sanatana Dharma and Vedanta. They translated it letter by letter but lacked spirit behind it. Sri Pujya R K Kashyap gives spiritual message for every shloka. The beauty of RV is; it has literal meaning, spiritual message and secret message. Secret message was passed on verbally to the Vedic disciple and now are lost. You can find secret message in RV on Gravity and Solar system. If you go by literal meaning, it means very ordinary but if you dig deep into it, it gives you science behind it.

The approximate date when RV was compiled varies from 1200 BC to 10000 BC. Hindus consider Vedas as Eternal and hence no date can be fixed.

RV is organized into 10 books known as Mandalas. Each mandala consists of hymns called Sukta (literally "well recited eulogy") Sukta in turn consist of individual stanza called rc (praise) which are further analysed into units of verse called Pada (foot). The meters most used in RV are Gayatri (three verse of eight syllables each i.e., total 24 syllables in three verses, Anushtub (4×8), Trishula (4×11) and jagati (4×12).

RV 10th Mandala 121 Sukta, Satapatha Brahmana (6.1.1.1-5) Taittarika Arankya (TA) and 10th Mandala 129 Sukta deal exclusively on Creation.

Literal meaning of X.121.1 of RV: (page 78). In the beginning arose golden seed (1); he was the sole lord of every creature. But what the poet wants to convey is that Hiranya (gold); the common meaning is gold. But hi stands for hita, 'placed' or hidden and ranya means (delight) in many RV mantras. Hence hiranya is that in which the delight is hidden. It is the concrete image of the higher light, the gold of the Truth. Gold is a symbolic colour of the light of the sun. Kah: the pronoun 'who,' Prajapati. Many translations of line (4) read who is the diety we shall worship? Reflecting doubt. The sages had no such doubts. They were wonder struck at entire creation (RV.10.121.10) clearly state that Prajapati is the Creator. Prajapati "Lord of people" is a group of Hindu dieties presiding over procreation and protection of like thereby king of kings (Raja).

Conclusion

Prajapati was later associated with Brahma, then Vishnu and then Shiva. Still later Prajapati was associated with Trimurty i.e., Brahma, Vishnu and lord Shiva. This topic will be dealt when we discuss about atom and also first two stanzas of Soundarya Lahiri by AdiShankaracharya.

X.121.1 toX.1.21.9 The sages were not sure whom to worship for the creation but also say except Him (X.121.1 to X.21.9). That means they knew which God was the Creator RV (10.121.10). This mantra clearly states that Prajapati is the Creator who should be worshipped.

Except Shatapata Brahmana (6.1.1.15) no other book viz, Genesis in Bible and Sura in Quran talk about what existed before Creation. Shatapata Brahmana clearly states that "In the beginning the non existence (viz unmanifest) alone was there." Rig Veda treats the topic of Creation in very original way. Entire Creation is dealt in metaphysical way in RV (X.129) "Non-existence (asat) then was not, nor existence of (sat); of movement, nor space beyond. What covered overall and where or what was any resting place? What were the Waters? Fathomless abyss."

When we recall creation Sukta X.129 and also X.90 "The Supreme Reality is not a mere existence, immutable and featureless. It is supremely aware, it is a Concsiousness. And this Concsiousness is again not a mere awareness. It is dynamic, it is Power. When the Concsiousness as Power moves into action, creation ensues. First, the Truth (Satyam) behind the creation, formless, itself leading to Law of the working of that Truth (rtam). This self-determination of Truth is the seed of Creation and its Law lays down the lines of manifestation and governs its development.

The most famous philosophical hymn (RV X: 129.1), which goes on to say "Non-existence (asat) then was not, nor Existence (sat) neither the principle of movement, nor space there beyond. What covered all (avarivasc) and where were waters? Fathomless abyss RV X: 129.2 Then neither death (mrtyu) or life (amrta), not any sign (Praketa) of night or day. The One breathless breathed by intrinsic power (svadha). None other was nor aught there beyond.

RV X: 129.3 of RV goes on to say "Darkness hidden in the beginning was this all. This all was ocean without mental consciousness (apraketam). All is hidden in formless being owing to the fragmentation of consciousness. Out of it, one was born by greatness of energy. (RV10.129.4)" "In the beginning desire arose. The primal seed of mind that was the first. The seers of wisdom found out in non-existent that builds up the existence (sat). In the heart they found it by purposeful impulsion (pratishya) and by thought – mind."

RV X.129.5 says "Their ray was extended horizontally, there was something above, and there was something below. The seed, all might was intrinsic power below, the purpose above." Note that the Creation is impelled by intrinsic power from below, the God of creation in the station above pulls up the consciousness levels to manifest and establish truth everywhere.

RV X.129.6 "Who knows it right, who can here set it forth, whence was it born, whence poured." These gods (devah) are from its pouring forth, whence then it came to be, who knows?

This above mantra and the text are viewed by some translators as indicating scepticism since they pose questions. These questions only suggest the wonder. The Creation is wondrous that we cannot even think about the One (the Supreme) who made it possible.

RV X.129.7 "From what source did this Creation come into being (1). Or whether one appointed it or not (2). He who is over eye therefore in Supreme station (3), he knows indeed or knows not (4)."

Part [3] and [4] are very interesting. The usual translation done by Indologists is 'he knows indeed or he knows not.' They are happy to note that even the creator does not know all. Whereas translation by A.K.C. is "He knows and he knows not." The idea is that in every act, the outcome is not really fixed at all till the last second. The grace can act at the last minute. There is no limitation.

We have heard of adage that 'not a blade of grass moves without His consent.' But this statement does not state that everything is planned in advance. In every action, there are so many possibilities for its termination. Only the Creator decides which possibility will prevail. The Creator does not need to plan ahead. Thus both the statements 'he knows' and 'he knows notin advance' are true. He does not specify the way of conclusions of an action in advance, since such a specification limits his own power. By definition, the Supreme Person has no limitations. The truth is fixed, the truth in movement has to take different forms in different circumstances so one cannot say, 'path given by movement is fixed in every detail for all times.' Inspiration be heard only by a person who is in contact with divine.

The creation in Rig Veda X: 129 by Steffen Stennud:

There is no evidence to suggest that RV (10:129) is a later addition and not part of original. Steffen Stennud when he is narrating has not mentioned the exact sukta and slokas of 10th mandala and 129 sukta. What he is referring is from 10:126.6. His interpretation as copied from Max Muler's translation is as follows:

Who knows whence this great Creation sprang?

He from whom all this great creation came

Whether his will created or was mute.

The most high seer that is in highest heaven.

Who knows it – or perhaps even He knows not

The above poetry is beautiful because the composer is wonder struck at the Creation. He expresses his doubt not because he doesn't know. The Highest seer that is the highest heaven. This only means that the Supreme Being in highest heaven has created this universe. Who knows it or perhaps He knows not. Having known that Supreme Being is the Creator but still the poet expresses his doubt not because the poet does not know who the creator was but puts forth the question 'perhaps

He knows not.' You should not take this piece of verse in isolation but must be understood in totality. Steffen Stennud is not an Indologist noris his work the original translation. He does not know the meters of poetry in Vedas nor is he a Sanskrit scholar. He niether knows the Vedantic psyche.

No where in Rig Veda RV: 129.1 to 10:129.7, never ever there is mention of universe being created out of nothing.

Mantra 1 goes on to say that there was neither asat (not existent) nor sat (existence). There was no movement or any space. This was the state before the Creation which is incomprehensible to human mind.

Steffen Stennud has a problem in stating that how nothing could have existed before creation. He is referring to verse RV X: 129.1 which states that "Non Existence (asat) then there was not Existence of (sat)." If you connect this with RVX:5.7 it says 'He is sat and asat is the highest station is the birth of understanding in the lap of mother adih.' If you connect the previous versewith RV X:5.7 it says "He is sat and asat in the highest station" which means sat has a form and asat has no form for it is being formless. Connect this again to RV 1.62.7 which says "He parted twofold the lying in the same abode for all time. He cannot be attained by mere effort" Mandikya Upanishad also says "The Brahman cannot be obtained by instruction nor by intellect nor much hearing"(3.2.3)

Steffen Stennud has just read different translations of 10:129 including tanslation of RV by Sayana of Vijayanagara kingdom in 14th century CE and then passing judgement. RV clearly says sat is nothing but Brahman. He did exist. How can Steffen Stennud say that nothing existed?

Again he has problem with the ending statement about nobody knowing of the world's origin. He like other Indologists are happy to note that even the creator does not know "He knows and He knows not." This has to be understood in this way – The idea is that every act, the outcome is not really fixed at all till the last second. The grace can act at the last second. There is no limitation. The creator doesn't have to plan in advance.

The Creation described in RV is not a sudden event like Big Bang. It is a gradual process.

Steffen Stennud sees two plausible explanations: Either we misinterpret the content of the hymn completely, or it has been altered since its original writing. To this it can be said that Steffen Stennud, not having read entire RV has no understanding of RV. As far as alteration is concerned RV is in poetry which has its metric form. Altering meter of poetry is very difficult. Then he goes on to say that 3 poets had composed the poetry. It is preposterous.

The Creation of Universe-Brihadaranyaka Upanishad.

Introduction:

Brihadaranyaka Upanishad (Bri. Up) is one of the oldest Upanishads. The Br Up is the 10th in terms of Muktika or cannon of 108 upanishads. It is estimated to have been composed in 700 BCE. The Sanskrit language text is contained within Shatapatha Brahmana, which itself a part of Shukla Yaurveda. Bri. Up is a treatise on Atman Brahma vidya or Atma Vidya includes passages on metaphysics, ethics and a yearning for knowledge that influenced various Indian regions, ancient and medieval scholars such as those by Madhavacharya and Adishankaracharya. Bri. Up literally means "great forest Upanishad." Yajnarvalka is credited have penned Bri Up. Yagnavalka is a great scholar and was invited by no less person than Janaka (father of Sita, consort of Sri Rama). It is interesting to know the end. Yagnavalka after imparting Atma Gnana (knowledge of spirit)to Janaka, Janaka becomes enlightened and says that he wants to renounce his kingdom. Yagnavalka also becomes enlightened while teaching Janaka and renounces every material thing to become a sanyasi (monk). Bri. Up represents great seers and thinkers among both men and women. This Upahishad has four such luminaries. Yajnavalka,

Janaka, Gargi and Maithreyi. The last two being females. Bri. Up gives concept of man and woman. "He was not at all happy. Therefore people (still) are not happy when alone, He desired a mate. He parted this very body into two. From that brcame husband and wife." Therefore, said Yajnavalkya this (body) is one half of oneself, like one of the two halves of a split pea (Bri. Up 1.4.3). Woman is eaual half of the man unlikein Genesis 2 verse 21out of the rib of Adam He made women.Bri Up has contributed two of the seven or eight mahavakyas i.e., great sayings. One is 'Ayam atma Brahma' meaning Soul is Brahman (Bri. Up 4.4.5).And 'Aham Brahmasmi' meaning I am Brahman (God) Bri. Up 1.4.10

Sri Pujya Krinananda has dealt the subject of The Creation of Universe - Brihadana Upanishad very well. It is not only spiritual but mind provoking He made the Sankrit of Bri Up which is very difficult to understand, very easy. There are number of metaphors and symbolisms.

The sloka 1 mentioned in this text is a part of 1.4.6 of Bri Up. The first sentence of this passage says: Tad ha idam tarhiavyakratam asit' meaning in the beginning the whole universe was undifferentiated. Then it goes on to say 'tat nama rupabhyam eva vyakriyata, asau nama ayam idam rupa iti.' Meaning One (Brahman) differentiated into various things with particular name and particular form. It is only name and form that makes differentied from that which is non-differentiated. Then how does differentiation take place? To this 'Tad idam api etarhi nama rupabhyam eva vyukriyate, asau nama ayam idam rupa iti' meaning – even today it is all differentiated only through name and form.

The original state which mentioned is the same of that in RV X:129. The interesting part brought out by Sri Pujya Krishmananda is on the subject and object. When subject which is you and object which is Brahman became one then there is nothing remaining. The separation of object from subject is what makes differentiation. This subject has been dealt when we were dealing with the first mantra of the shloka ubder reference.

Conclusion

Referring to Satan by Sri Pujya Krishmananda he is referring to Lucifer – Literally meaning shinning One, also Light bearer of Sun of Morning. He is also known as Satan or the devil. The name of Lucifer appears only once in Bible (Isaiah and falling of Satan in 14:12-20). Apparently Satan is the name of the Devil and he is rebellion. Pujya Krishnananda had to bring falling of Lucifer to show that, man when he stopped looking inwards to know Atman started to look for Brahman high up in the sky. Atman and Brahman are synonyms.

Lucifer in Isiah 14:12-17 goes on like this

12. How you are fallen from heaven,
    O Lucifer, [a] son of the morning!
    How you are cut down to the ground,
    You have weakened the nations !

13. For you have said in your heart:
    I will ascend into heaven,
    I will exalt my throne above the stars of God;
    I will also sit on the mount of congregation
    On the farthest sides of the north.

14. I will ascend above heights of the clouds,
    I will be like the most High
    Yet you shall be brought down to Sheol,
    To the lowest depths

15. Yet you shall be brought down to shed
    To the lowest depths of the Pit

16. Those who see you will gaze at you
    And consider saying you, saying
    Is this the man who made the earth tremble
    Who shook kingdoms

17. Who made the world as wilderness
    And destroyed all the cities
    Who did not open the house of his prisoners?

18. All the kings of the nations
    All of them sleep in glory
    Everyone in his own house

19. But you are all cast out of the graves
    Like an abominable branch
    Like garment of those who are slain

20. You will not be joined in burial
    Because you have destroyed your land
    And slain your people
    The brood of evil doers will not be named.

Aitereya Upanishad

The Aitereya Upanishad (Ait. Up) belongs to the Aitereya Aranyaka and is part of the RV. Considering the three parts, the Upanishad proper brings with fourth section of the second part of Aranyaka and comprises section four, five and six.

The Upanishad as already noted is divided into three parts: The method by which self knowledge is taught is twofold.

The first part of the Upanishad consists of three chapters, with a total twenty-three verses and describes Creation. Ait. Up is a primary ancient Upanishad and is listed eight in Muktika canon of 108 Upanishads. Considered one of the middle Upanishads, the date of composition is not known but has been estimated by scholars to be sometime around 6th or 5th century BCE. Ait. Up also brought out one of 7 Mahavakyas "Pragnanam Brahma"(Ait. Up 3.13) meaning Brahman (God) is Consciousness. Late Sri Pujya T N Sethumadhavan through

Ait. Up brings out a interesting point saying that Ait. Up is a forerunner of 'unified Field Theory' or a Theory of everything which the modern phycisists are trying to discover. Many of the modern day scientists of Quantum physics find answers in Upanishad. Citing examples are – The famous Danish physicist and Noble Prize winner, Laureate Niel Bohr (1886-1962) was a follower of vedas. He said 'I go into Upanishads to ask questions' Both Bohr and Schrodlinger, the founder of Quantum physics were avid readers of the Vedic texts and observed their experiments in quantum physics were consistent with what they had read in Vedas.

Erwin Shrodinger said 'The Multiplicity is only apparent. This is the doctrine of the Upanishads only. The mystical experience of the union with God regularly leads to this view unless strong prejudices stand in the West' P 129 Cambridge University press by Erwin Shrodinger.

The question of Creation of Universe is not only scientific but also philosophical and spiritual.

Like every Upanishad Ait Up also begins with mantra. The mantra is not on any personal God but on Brahman (the one and without the second, ever expanding, consciousness, ultimate infinite Bliss – there are many others attributes which will not be dealt here).

Manvantara theory by Dr. SL Dhani

Punch lines of Manvantara theory

1. Brahma, Vishnu and Rudra (Shiva) are represented as cosmic eras and not as Gods.
2. Precess of creation begins with day of Brahma and that of dissolution as the night of Brahma.
3. One day and night of Brahma is equal to duration of 4320 million years each.
4. Brahmanda is a sphere.
5. Life span of Brahma is 100 yrs each year being 360 days (the day includes night). The life of Brahma is thus 8640 million ordinary years.

6. This period is equal to only a day of Vishnu who is similarly has a full life of hundred years which in turn is equal to a day of Rudra (Shiva).
7. The present cycle of creation started, according to Puranas 1972949076 or about 1973 million years ago.
8. We are in 7th Manvantara, 27 Caturyuga have elapsed and at present we are passing through the Kaliyuga of 28th Caturyuga of the said Vaivasavata manvantara.
9. Vaivasvata Manvantara started 120 million years before present.

Kala (Time) is a fundamental basis of creation. It cannot be pinned down. It is powerful, ineffable dimension that holds the entire universe together. This is called Kala. The aspect of time is a result of cyclical movement of physical reality. From atomic to cosmic, everything physical is in cyclical motion only in time space is possible. Hence space is a consequence of time. Because of space form is possible. Because of form, all physical reality becomes possible. Even gravity is one small by-product of time.

Biblical Doctrine of creation of God by Rev Dr. G. Wright Doyle

Introduction: Creation of Universe and man is given in Book of Genesis. Although Genesis does not directly name its author and although Genesis ends some three centuries before Moses was born, the whole scripture and church history is unified in their adherence to Mosaic authorship of Genesis.

"God is spirit not a field of energy or external matter. Thus he is invisible. No one has ever seen his essence" as quoted by Rev Dr. G W Doyle.

Kena Upanishad (1.3-) also states:

"The eye does not go thither, or speech, or the mind. We do not know It."

Conclusion

Svethaswetara Upanishad (1V.20) "His form does not stand within the range of the senses. No one perceives Him with eye."

"God is unitary" quote from Rev Dr. G. Wright Doyle."

"Ekameka Adhvatiya Brahman" meaning God is one without second.

"God created world out of nothing."

Samkhya Sutra aphorism "From nothing nothing can be created."

"With God for with God all things are possible Mark 10:26-27.

RV (10:129) also says that there was nothing in the universe before creation except darkness, hunger and death.

Even RV, Upanishads and Manvantara theory of Creation, the creation did not occur suddenly but step by step as stated.

Genesis God created this earth in six days Genesis(I:1-27). God creates the heaven, earth living creatures according to its kind, cattle, creating things and beast of earth, each according to its kind, and it was so. These all things He created in six days. Then God thinks of creating Man and Woman, "on 7th day He takes rest" Genesis 2:2.

Genesis chapter (1:26-27)

1:26 The God said "Let Us make man and in our image, according to our likeness; let them have dominion over the fish of the sea, over the birds or the air and over cattle, over all the earth creeping thing that creeps on earth"

So God Created Man in His own image, in the image of God He created him, male and female He created them (1:27).

Chapter 2 verse 7 of Genesis "And the Lord formed Man from the dust of the ground and" breathed into nostrils the breath of life and man became living being. Ait up also says similarly.

1:2:18 And Lord God said "It is not good that man should be alone. I will make him a helper comparable to him.

1:2:21 And the Lord God caused a deep sleep to fall on Adam, and he slept. He took out one of his ribs and closed up the flesh in place.

1:2:22 Then the rib which Lord God had taken from man He made into woman and He brought her to man.

Creation of man in Genesis is conflicting:

1:27 So God created man in His own image, in the image of God He created him and female. He created them. But again in chapter 2 verse 7 Lord God formed man of the dust of ground and He breathed into nostrils breath of life.

He created woman out of rib of the man and names her woman (1:22).

The contradiction is, first God created man in his own image and female He created them according. This means God created Man and Woman twice and not once if we were to go by 1:27 and 2.7 mentioned above. What happened to his first creation where He made man in His own image and woman.

Genesis 4.1 Now Adam knew Eve his wife and she conceived and bore Cain and said "I have gotten a man from the Lord" (4:1).

Then she bore again this time his brother Abel (4:2)

4:8 Now Cain talked with Abel his brother and it came to pass were in the field that Cain rose against Abel his brother and killed him. When God comes to know that Abel was killed by Cain (4:8)

"So now you are cursed from the earth, which has opened its mouth to receive your brother's blood from your hand (4:11)"When you till the ground, it shall no longer yield its strength to you. A fugitive and vagabond you shall be on the earth" (4:12). And Cain said to the Lord "My punishment is greater than I ever bear !(4:13)

4.14 Surely you have driven me out this day from the face of the ground. "I shall be hidden from your face. I shall be fugitive and a

Conclusion

vagabond on earth and it will happen that someone who finds me will kill me."(4:14)

From the above passage 4.14 Cain tells God that he may be killed. Who would be those who would kill Cain. Now there are only three persons as per Bible, 1 Adam 2 Eve and 3 Cain.

Then why did Cain have to say that he would be found and killed. This means there are already humans/humanoids which existed on earth otherwise who would kill Abel.

4.17 And Cain knew his wife and she conceived and bore Enoch. This means Cain marries and has a son through his wife. If Adam, Eve and Cain are only there on the earth then whom did Cain knewas his wife and bore a son? This means that women were also present on earth besides Adam, Eve and Cain.

As per Genealogies in Genesis 5:1.32 and 9.28 10.32 God created world about 6000 years ago (i.e., as on today) Psalms 90:4 "For a thousand years in your sight are like yesterday when it passed." If we calculate the reference in psalm 90:4 and (Genesis 5:1.32 and 9.28 10.32) then 1000 X 6000 will be 6000,000. That makes 6 million or 60 lakh years.

Genesis 6:1 "Now it came to pass, when men began to multiply on the face of world daughters were born to them." (Genesis 6:2) That the sons of God saw daughters of men, that they were beautiful; and they took to wives for themselves (6:2) of all the chose i(6:3) of Genesis. Does it mean that aliens were marrying women of earth who were beautiful.

God Created this universe to reflect his own beauty and goodness.

Bri Up II-iii-1: Brahman has two forms gross and subtle. The gross is the manifested universe.

In the Book of Genesis the cause of Mankind destruction is given "And God saw that the wickedness of man was great in the earth and every imagination of thoughts of heart only evil continually/and it repented the Lord that he had made man on the earth and it grieved him at his heart. And the Lord said," I will destroy man whom I have created from the face of the earth, both man and beast and creeping thing and the fowls of air, for it repenteth me thatI have made them."

In the story of Manu, however, the destruction of world treated as a part of natural order of things, rather than divine punishment.

Manu was said to have three sons before the flood – Charma, Sharma and Yapeti, while Noah also had three sons – Ham, Shem and Japheth.

Abrahamic God believed by Jews, Christians and Islam depict God in Masculine gender, while in Upanishads and Vedas no gender of God is mentioned.

Sevatasvetara Upanishad 4.3 "You are the woman, you are the man, you are the boy and you are the girl too. You are the old man tottering with a stick. Taking birth, you have your faces everwhere."

Quran

"Have the disbelieves not seen the skies (space) and the earth (matter) were joined to together then we ripped them apart?" (21:30) The above quote from Quran does not say anything about singularity before creation or existence prior in the beginning Ref RV (X:129) for the state prior to Creation.

"Angels and the Ruh ascend up to Him in a day, the measure of which was 50 thousand years." (70:4)

"A day with your Lord is like thousand years of your count" (Quran 22:4). Psalms 90:4 also says the same that 1000 man years one year of God.

As per Quran based on 70:4 and 22.47 the age of universe should be 18.25 billion years.

Conclusion

Quran on multiverse

"Did you not see that God created seven universes in layers?(Quran 71:1)

The barriers that exist between these universes are mentioned in Quran.

"If you are able to penetrate through the regions of heaven and the earth, then go ahead and penetrate; you cannot penetrate without authority"(Quran 55:33).

Existence of multiverse is not new to Hindus for there are a number of references in Srimad Bhagwatam (SB).

Big Bang Theory

Big bang(BB) theory is based on explosion from a single entity with infinite mass in zero space. It occurred in 10^{-37} seconds and ever since it is expanding.

Brief points against Big Bang theory (BB)

1. Mature galaxies exist where the BB predicts infant galaxies (like the 13.4 billion years distant BB prediction).
2. An entire universe – worth missing antimatter contradicts most fundamental BB prediction.
3. Observations show that spiral galaxies are missing millions of years before BB predicted collusion.
4. Cluster of galaxy exist at a great distances where the BB predicts should not exist.
5. A trillion stars are missing an unimaginably massive quantity of heavy elements a total of nine billion years worth.
6. Galaxy super clusters exist yet the BB predicts that gravity could not form them even in the alleged age of cosmos.
7. A missing generation of billion first stars that failed search has implied simply never existed.
8. Missing uniform distribution of earth's radio-activity.
9. Solar system formation theory wrong.

10. It is 'philosophy' not a science that make BB claim that the universe has no centre.
11. Amassing evidence suggests that universe may have a centre.
12. Sun is missing nearly 100% of spin that natural formation would impart.
13. The beloved supernova chemical evolution story for formation of heavy elements is now widely rejected.
14. Missing uniform distribution of solar system isotopes.
15. Missing billions of years additional clustering of nearby galaxies.
16. Surface brightness of the furthest galaxies against a fundamental BB claim is identical to that of nearest galaxies.
17. Missing shadow of the BB with long – predicted 'quieter' echo behind nearby galaxy cluster now disproved.
18. There should exist a supermassive black hole an iron poor star and a dusty galaxy – but they also not there.
19. Fine tuning and dozens of other MAJOR scientific observations and 1000+ scientists doubting BBObservation made by NASA

1) According to Vedic scripture, universe has existed for 155.522 trillion years. Before this cycle there were countless other cycles which will end in 155.8 trillion years time.
2) The cosmic creations 155 trillion years old – 1000 of Brahma mean 1000 × 2 × 155 trillion years of creation have passed. Brahma creates universe. Vedas say that thousands of Brahmas have passed away! In other words this is not the first time universe has been created.
3) The local streaming motions of galaxies are too high for a finit universe that is supposed to be everywhere uniform.

Invisible dark matter of unknown but non-baryonic nature must be dominant ingredient of entire universe. (This dark matter is nothing but naval of Lord Vishnu from where birth of brahma takes place) Source Facebook dated May 22, 2016

No Big Bang? Quantum equation predicts universe has no beginning

Quantum equation tries to prove that universe has no beginning. Most of the material found in this topic is to negate Einstein's General Relativity.

The theory put forward by Quantum physics that the universe has no beginning. This goes against Manvantara theory of Evolution. Creation and destruction of universe is cyclical and takes place n number of times.

Top Ten Mind Boggling theories – Toptenz.com

Punch lines: In our solar system we have eight planets but we still don't know what is beyond Pluto and still within our solar system, meaning there could quite possibly be more planets that we haven't yet discovered.

The first evidence of another planet within our solar system first arose after the discovery of two possible dwarf planets – 2012 VP113 and 90377 sedna. Something massive is affecting the orbit of these two possible dwarf planets. It is believed that there is a planet 10 times bigger than earth that is affecting their orbit.

These planets are at least 200 astronomical units away from earth. One astronomical unit is distance between the sun and the earth, which is about 93 million miles.

White holes are also hypothesized. If black holes are entry points then white holes are exit points. If one comes out of white hole and end up in other part of space or in a completely, different time once they come out of white holes. This is only speculative Quantum Entanglement where tiny particles, like electrons that were formerly, interact with each other even after they are separated. The interesting thing is that if you change one particle, it also changes the one formerly entangled even if the two particles are a galaxy apart. That means it happens at a speed higher than Einstein's Theory of Relativity.

Physicists believe that in the fraction of a second after BB it picked up power before tapering off. This fraction of second is called 'inflation.' Where as RV, Upanishad, Bible, Quran state that creation was a graded process.

A mind boggling Question is what sparked the Big Bang? According to Ekypyrotic universe theory by Paul Steinhardt, it was cause by collusion of two three-dimensional worlds.

Multiverse theory is not new to Hindus because there are a number of shlokas in Srimad Bhagwatam saying that there are innumerable universes.

A Universe from nothing?

Some scientists feel universe is eternal without beginning. But vast majority who accepted BB say universe has a beginning. There is something called "virtual creation" and annihilation. There are number of subatomic particle which exist only momentarily but can cause effect which can be detected. Short lifetimes of these virtual particles are governed by Heisenberg Uncertainity Principle (HUP) which says that a short lived state cannot have a well defined energy. The spontaneous (but short lived) subatomic particles from a vacuum are called quantum fluctuation.

Krauss et al evolutionary physicists argue that universe itself the result of quantum fluctuation. This speculation is found to be baseless.

Evolutionary physicists argue that if total energy content of universe were exactly zero, then a universe resulting from such a fluctuation could persist indefinitely without violating the HUP.

Stephen Hawking writes that "The idea of inflation could explain why there is so much matter in the universe" – The answer is that, in quantum theory, particles can be created out of energy in form of particle/antiparticle pair. But that just raises the question of where the energy came from."

Conclusion

Saudarya lahari by Adishankara has an answer for this:

1. Lord Shiva only becomes ableto do Creation along with Shakti (energ)

    Without her Even an inch cannot move

    And so how can one does not do good deeds

    Or one who does not sing your praise

    And so how can, one who does not sing Your praise,

    Become adequate to worship you

    Oh, goddess mine who is worshipped by trinity.

    The great Adisesha with is thousand heads some how carries the dust of your feet, With great effort carries a dust of your feet.

2. Lord Brahma, the creator of yore, selects a dust from your feet, and createts he this the world.

    Great Adisesha with his thousand heads,

    Some how carries dust of your feet

    With great effort Lord Rudra,

    Takes it and poudered it nice

    And uses it as Holy ash.

    Shakti (ENERGY) provides energy of an atom. Dust of Shakti's feet is described as smallest particles within the nucleus of Atom.

    Lord Brahma is the creator depited as electon, Vishu as Proton and Shiva as Neutron.

The answer is that the total energy of universe is exactly zero. According to Vishnu Sahastranama which contain 1000 names of Lord Vishu(one of Hindu trinity), Vishnu is also called SHUNYA meaning cipher/zero. Since God is immeasurable it seems plainly alright to name him "ananta" the infinite. But how is one to explain hailing almighty as Shunya, the cipher. There is a view that if infinity is immeasureable, so is zero. Mathematically speaking one could define zero to be anti-infinity.

If infinity is immeasureable platitude so is zero. Mathematically speaking one could define zero as immeasureable series of values from zero to infinity floating somewhere out there is endless space, then surely zero would be at one end of it, while infinity would be found at the other end. And if you reflect upon deeply, that would make out "zero" and "infinity" to be two sides of the same ungraspable coin. In his commentary AdiShankaracharya (600 CE) explains "shunya" as an apt name for God the supreme Brahman, According to AdiShankara school of metaphysics God is "Gunashunyam" MEANING ZERO GUNAS such as Sathvic, Rajas and Tamas.

God by corollary is also totally devoid of inauspicious, un-wholesome or negative qualities. God is also nirguna i.e., being totally devoid of any quality, at the same time Brahman is replete with infinite good attributes. According to Bhattar He is absolutely bereft of defective qualities and He is to be known as God of "zero defects" – in other words Shunya.

It is not possible to know that total energy of universe is zero. Krauss urges the quantum gravity (a theory that merges quantum mechanics and general relativity). But quantum gravity does not yet exist.

The laws of physics simply do not apply when the universe popup. Glaring fallacy of argument: 'While the laws of physics and chemistry in our universe do indeed allow life to exist, then do not allow life to evolve.' The law of physics and chemistry simply are not favourable to the evolution of life.

Multiverse:

Buddhism and Hindu texts mention about existemce of multiverse. Srimad Bhagwatam (SB) gives ample evidence to it in number of verses. To site one example SB X 1:16:40 states "Brahmanda (cosmic shell) which consists of eight special modifications and sixteen effects of Prakriti, has dimension of fifty million Yojanas (9 miles make one yojana) within and is surrounded with layers of the five elements, the layer of each element being ten times the distance of internal dimension of preceeding shell. All this is only like an atomic particle in Sri Hari (Vishnu). In Him there are countless other shells like this, all of which together even is like a few atomic particles for Him." Brahma Vaivarta purana also gives evidence of parallel universe.

Is the Universe Conscious?

To most people the idea that the universe might be viewed as a conscious entity is outlandish in extreme. Mind is evolved from brain which is in turn a product of evolution. Human body is composed of two things – gross one, the body and subtle one, the mind. Mind is not something separate from matter, it is a process of embodied in matter.

The physical universe ceases to be an unconscious object, observed and explored by Conscious minds which somehow stand above and outside. Consciousness of universe; our Consciousness is a part of the universe. As Carl Sagan puts it, 'humans are the stuff of cosmos examining itself.'

'Ekam sad Vipra Bahudha Vedanti' (RV: I: 164-146) Brahman (God) alone exists, sages call it by various names viz, Brahman, Yehva, God the father, Jesus and Allah. 'God is Truth' Thai. Up (Thai.Up,)' I and the way and

Truth'(John 14:6) and Al-Haqq one of the names of Allah meaning The truth. Hence if the God is The Truth in all major religions then why do we fight? Holy (Quran).Quran has given 99 names of Allah like Al Quddus meaning Pure one, Al latif, means The Subtle one, Ash Shahid means The witness, Al Haqq means Truth, Al Batim – The Hidden one, Al Zahir – The magnificient one, Al Nur – The light, Al Baqui – The Everlasting one, so on and so forth. The names only indicate attributes of God. Such references can also be found in Vedas, Upanishad and Holy Bible.

Plato in his Parenides, where he treats sublimely nature of God, says "nothing can express his nature therefore no name can be attributed to Him. That is why Sanatana Dharma (Hinduism) called It, Brahman meaning that which expands. In Exodus when Moses asks God for His name He says 'I AM WHO I AM' God here doesn't want to be named because so many names about 1300 names already existing in Egypt whom Pharos were worshipping. I God of Moses has told His name then His name would have been one I addition to 1300 names of gods worshipped by Pheros Hence when Moses asked for His name that is what He had said "I AM HE WHO I AM." The moment you name, you associate it with form.

Now let us see what Universe really contain of. Ordinary matter which includes stars and galaxies account 4.9. % of content of universe, dark matter and dark energy. The present overall density of this type of matter is very low roughly 4.5×10^{-3} gms for cubic meter of volume.[129]

The universe is composed completely of dark matter, dark energy and ordinary matter – other contents are electromagnetic and antimatter[130] and dark matter. All these are homogenously distributed throughout universe over scale longer than 300 million light years or so.

The observable universe contains approximately 300 sextillion[131] (3×10^{23}) stars.

And more than 100 million galaxies.[132] Most of the universe is emptiness, the largest known void measures 1.8 billion light years.[133] The stars in the Universe are are all made of atoms. The volume of an atom (roughly 1 angstron or 10^{10} m in diameter) is about 15 orders of magnitude larger than value of anucleon (roughly 1 f m or 10^{-13} m) in

---

[129] NASA/WWMAP science team (29 Jan 2014) universe 10.1. What is universe made of

[130] A Fritzsche, Hellmutt, 'Electromagnetic radiation/Physics, Encyclopedia Brittanica P.1 Retrieved 2015-07-26.

[131] Mandolesi N., Carlzolari P., Cortiglionis N, Delphino Fet etc "Large scale homogeneity of Universe measured by microwave background" Nature 319 (6065) 751-753

[132] A Meckie Glen (Feb 1, 2002) To see the universe in grain of Taranaski Sand Swimburne University Retrieved 2006-12-20

[133] Astromer discover largest known structure in universe is – a big hole. The Gaurdian 20 April 2015.

Conclusion

diameter. That means atom is 99.9999999999% empty. Hence most part of universe is emptiness. According to Sadguru of Isha blog defines Shiva as "That which is not", meaning void.

Theory of Creation of Multiverse Universe as per Bhagwat Gita: There are millions of Universes with millions of planets. All universes have life, are closed of different size and properties. The whole material creation with millions of unverses constitute just a quarter of creation.

One day of Brahma is 4-32 Billion years. We are currently half way macking approximately 2.16 Billion years in the current cycle. Static universe model fit observational data better than expanding universe model. The microwave background makes more sense as the limiting temp of space heated by starlight than as the remanant of a fireball.

The universe has too much large scale structure (interspaced "walls" and void) to form in time as short as 10-20 billion years.

The average luminosity of quasars must decrease as short as 10-20 billion years. The average luminosity of quasars must decrease with time in just the right way so that their average apparent brightness is same at all red shifts.

The ages of globular clusters appear older than universe. Darkness literally the absence of light. Light emitted from photon but darkness has no such particle like light. Darkness is just absence of electromagnetic wave (light).

If you define speed of dark to be speed of light at which you find out about same one as turning the lights off i.e., 971 million mph or 2.99×10^8 m/s.

It takes 8 minutes for light to travel from sun to earth. For e.g., if sun wastoo suddenly get lit out, and then darkness wouldn't get to earth in 8 min meaning darkness would be immediate as soon as sun is lit off. Meaning Darkness is faster than light. Most of thr Universe is dark.

Every one thinks we need to be able to travel faster than speed of light to go through space. In actuality, we need not to be able to surpass speed of dark. Darkness is constant. Turn off the stars darkness prevails. Darkness is not absence of light because darkness exists even in when light is present. There is a saying in Hindi 'Diya tale andhera' meaning there is darkness underneath the lamp. Actually light is invisible. We see light which is reflected on an object. Light is one which humans see that too in limited spectrum, light and darkness coexist.

At fundamental level everything is energy. Early in 20th century the unquestioned assumption that physical universe is actually physical head to a scientific search for elementary "point particle." Upon which all life is built which would prove that reality –was not an illusion. But as soon as scientists began smashing electrons and other particles in enormous accelerators, they quickly realized the foundation of physical world weren't physical at all – that everything is energy.

Quantum Weirdness: The evidence that everything is energy.

The solar system picture of electrons and protons as tiny solid, planet like structures whizzing around a large neutron, mucons, tauons, quarks and gluons have no internal structure and no physical size, meaning that they are entirely illusionary as one puts another way, made up of energy. They are zero dimentional and more like events than things.

As if that were not bad enough electrons (these negatively charged particles that aren't really particles) were discovered to be both wave and a particle at the same time (wave particle duality). Electron showed up in one form or the other depending on the experiment involved. If they had to pinpoint – after all, when everything is energy, it is hard to keep in one place.

Scientists can know a particle's velocity or its position, but not both at the same time. Another strange habit energetic particles have is they can be in more than one place at the same time called "Superimposition." Electron and other non-particle particles are capable of being in hundred places simultaneously, which is only possible if everything is energy at the most fundamental level. In fact, towards the end of his life, when Einstein was asked what was the biggest physics question he wished. "I'd be happy if I just knew what an electron really was."

Reality is stranger than we think:It is strange that most people can think. And the reality is that everything is energy and energy is everythin. Quantum physics has amazing things to say about energetic nature of reality that can free our minds from superimposition and restrictions. Wave-particle duality casts doubt upon the very foundation of scientific method itself. Objectivity and necessary separation of scientists from experiment. Isn't objectivity the Holy Grail of science? Yes it is. But at subatomic level interactions and observations have been shown to effect and even determine experimental outcomes, which ultimately points to the possibility that there is no such thingas: separation and objectivity.

Everything is energy, it responds to Consciousness itself. In other words, it is possible that unless some agency (such as human consciousness) interferes, particles remain in probabilistic energy-wave state and never actualize into one location in particular form at all. Ultimately, the reality as we experience it seems to be result of human consciousness inferring that with the quantum level of existence that we are pure wave energy since all of us are made of atoms.

We tend to think that everything is real but actually only information that the brain receives and translate into picture called reality. Amazingly scientists are beginning to think same way that understanding energy is everything. Energy is information in its purest form.

The Copenhagan Interpretation isn't the only indicator that an information based matrix of energetic reality is what we are dealing with. Entanglement is another freaky physics. Conunundrum pointing that way. Once particles have interacted they become entangled which means forever they effect each other spin (which isn't a spinning motion at all but something called angular momentum). They are connected by unseen energy of force that permeates everything.

It is as if once two particles have kissed they become life long penpals. No matter how far apart they get, if scientists change the spin state of one entangled electron it is guaranteered its partner's spin state will change in opposite direction in response every time instantaneously. Even if they are million light years apart. This can only happen if everything is energetically connected on same level. Which means either we ignore Einstein's theory of special relativity and its prohibition bearing signal and accept that particles are somehow breaking the speed limit and communication instantaneously across vast distances at subatomic level.

Replicated studies show that living cells can instantly communicate over distances. One of the simplest experiments involves a batch of algae cells grows in a petri dish. After few days the cells are dividing and the rest of the cells are whisked away to a different lab. When the original group of cells is simulated by low voltage current, the separated group of cells in the lab miles away react in precisely same way.

Noted English mathematical physicist Sir Roger Penrose theorizes that at the Plank scale (an unfathomly small and unimaginally energetic scale at which even quantum field breaksdown) the entire universe is actually pure, abstract information. Plank is the smallest subatomic particle, when compared neutron is larger. Plank lenth is $1.616229(38) \times 10^{-35}$ meters

Illusion of Reality: Science find that everything is energy and Reality isn't what we think is.

(source: The illusion of reality. Scientists find everything and reality isn't what we think Jan 27, 2017. Google search)

Have you ever thought of what is really real? Well the latest discoveries in quantum physics show that nothing is as it seems. In fact, it has been proven that everything around us is made of pure energy.

Many experiments have been made, and with new advancement in quantum physics, more and more information points to the fact that at core of all existence lies a pure energetic network that seem to interwine in way that construct our reality.

As soon as scientists started smashing electrons in enormous accelerators, they found that building blocks are all matter is in fact energy. Depending on circumstances, these subatomic particles behaved differently. Interaction and observation were shown to affect the way these particles behave.

Simple experiment in quantum physics is passing photons in double slit. In observed result you find two verticle double lines while in unobserved you find several bands cancelling each other. Perhaps the world around us exists simply because of a kind of Consciousness holding it together into existence.

This Consciousness has been referred to as 'collective consciousness,' 'God,' 'Logos' and many other interpretations which posed theory that all existence is at it is because of some kind of consciousness holding it together. "Vasudaiva kutumbam" Maha Upanishad Ch VI 71-73 meaning, the world is one family. This is the greatest message Maha Upanishad has given to the world. We are all connected to each other at consciousness level. "Actually your soul and mine are same; we appear and disappear in each other." Rumi.

All these while we were talking of energy and consciousness. Upanishads are professing all the time about consciousness. Aitereya Upanishad 3.1.3 "Prajnanam Brahma" meaning Brahman is consciousness. This universe is nothing but manifested Brahman (God) and hence Universe is Consciousness.

"Sarvam Kalv idam Brahma" Chandogya Upanishad 3.14.1 meaning 'All this that we see in world is Brahman.' Considering Brahman is Consciousness. Ait up 3.1.3 Brahman is in reality, knowledge (Al Haqq is one of the names of Allah which means knowledge). Knowing that we are all in essence made up of same elements that were present during creation of universe it seems that everything around us, everyone around us, including us – everything is connected.

As Swami Muktananda puts it "The Self Shiva is supremely pure and independent and you can experience it constantly sparkling within your mind. It cannot be perceived by senses, because it makes the senses function. It cannot be perceived by mind because it makes mind think. Still the self can be known and to know it we do not need the help of mind or the senses."

Katha Upanishad (3:7)

"Higher than senses are the objects of sense,

Higher than the object of sense is mind,

Higher than mind is intellect,

Higher than intellect is Self (Atman, soul),

Higher than self is the unmanifest

Higher than the unmanifest is the Supreme personified

Highest is the Supreme the God ultimate"

Bha. Gita 3:52 "The working senses are superior to dull matter; mind is higher than the senses; intelligence is still higher than the mind; and he [the soul] is even higher than intelligence."

R C Henry Professor of physics and astronomy at John Hopkins University.

"A fundamental conclusion of new physics, acknowledge that observer creates the reality." As observers, we are personally involved with creation of our own reality. Physicists are being forced to admit that the universe is a 'mental' construction. Pioneering physicist Sir James Jeans wrote "The stream of knowledge is heading toward a non-mechnical reality; the universe begins to look more like a great thought than like a great machine. Mind no longer appears to be an accidental intruder into realm of matter; we ought to rather hail it as the creator and governor of realm of matter. Get it over and accept the inarguable conclusion. The universe is immaterial – mental and spiritual."

"Observation not only disturbs what has to be measured, they produce it. We compel the electron to assume a definite position. We ourselves produce the result of measurement (source http://deanradin.com/evidence/Rodin2012doubleslit.pdf)

In USA experiments are being carried out on psychokinesis experiment, the global consciousness experiment, intelligence agency remote viewing experiment, thoughts and intentions altering the structure of water, the placebo effect, teleportation studies and many more.

In Tulsi Das in 'Ramacharitha Manas' a book written in Awadhi Hindi translated from ValmikiRamayana. Ramayana occurred in Tretayug while Tulsi Das lived in 15th Century near Varanasi.

In this Sita consort of Lord Rama gives boon to Hanumana. The couplet (Doha) goes like this

"Ashta Siddhi Nau Nidhi ke Data

As-vara Deen Janaki Mata"

Janaki is another name of Sita-consort of Lord Rama.

The couplet means that Sita gives boon of 8 supernatural powers to Hanumana who was a true disciple of Lord Rama. The Boons are as follows:

1) Anima: Reducing one's self to the size of an atom
2) Mahima: Growing one's physical self to incredibly large size
3) Garima: Making one's body so heavy that nothing can move
4) Laghima: Becoming almost weightless
5) Prapti: Being able to go/travel anywhere one wants
6) Prakamya: Being able to obtain whatever he wants
7) Istva: Possessing lordship
8) Vastna: Being able to control other's mind

These supernatural boons may be bestowed upon by God or can be obtained by yogic meditation under a guru. Svetasvetara Upanishad

1:4 "the eight supernatural powers, endowed with which man can make himself small as an atom, huge as a mountain, light as air, can reach any objecthe likes, rule everything he wants, conquer everything, and fulfil all desires."

Whenever you offer anything to a guest, offer with good intension and also a blessing. Buddhist monks bless tea with good intension. Proof that Consciousness changes the molecular structure of water is proved by Masaru Emoto.

Masaru Emoto a Japanese Researcher studied the physical effects of words, songs and prayers. Photos show how molecular structure of water changes after being exposed to different sounds.

Positive messages produce more pleasing while negative stimuli produce less coherent forms. The average human body is 60% water. You can see for yourself how positive messages can influence our own physical bodies.

Plate 1

Heavy metal Sound **Amazing grace**

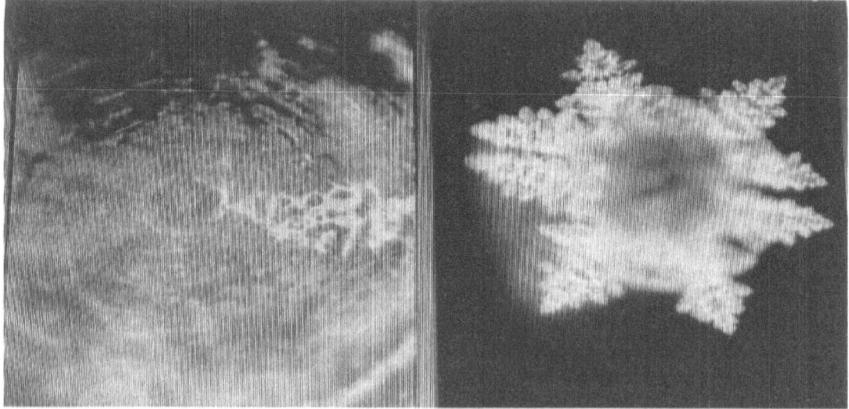

Conclusion

Evil   Love

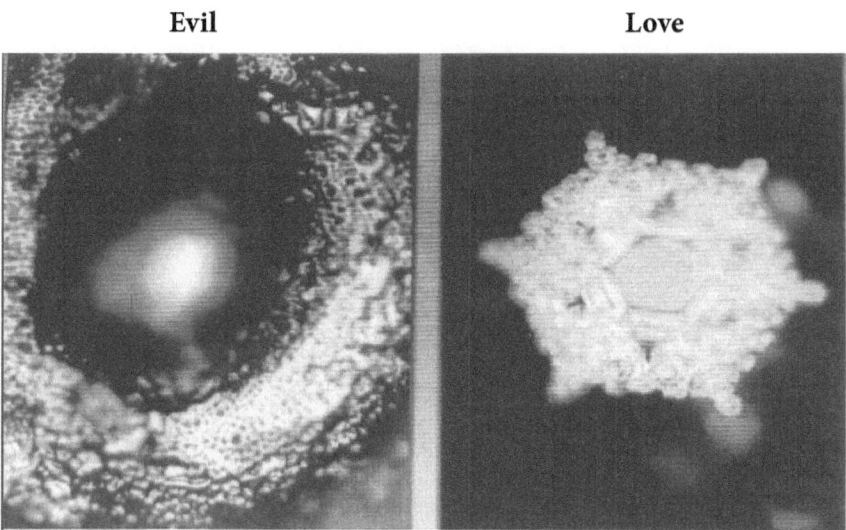

"Thank you"   "You fool"

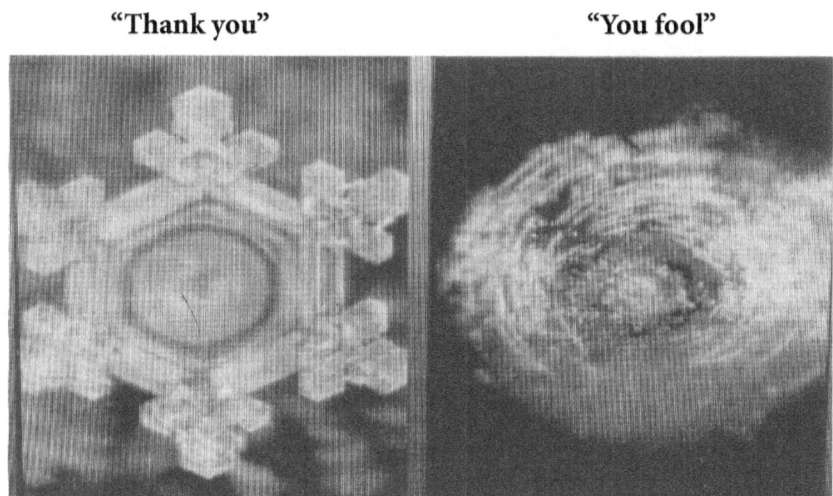

Plate 3

| Before Buddha Prayer | After Buddha Prayer |

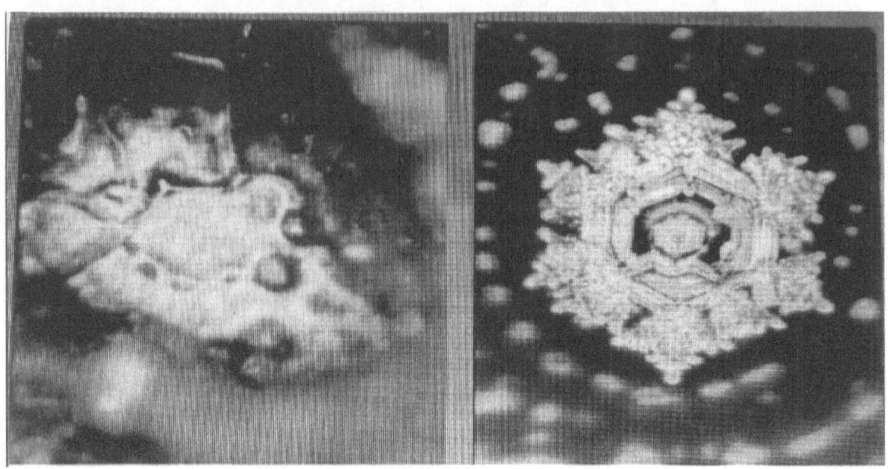

Plate 4

There is great power in prayers too. Eleventh Anuvakaof Sri Rudram and Purusha sutkam has great healing power when recited. A great example of quantum physics meeting ancient wisdom is in fact that Nikola Tesla was influenced by Vedic philosophy when pondering his ideas of zero point energy. Nikola Tesla said "the day science begins to study non-physical phenomenon, it will make more progress in one decade than all previous centuries of its existence. To understand true nature, one must think it in terms of energy, frequency and vibration."

Power of sound:

Fotrs of Jerico fell down on blowing of trupets (Jashua 6.1-27). Such is the power of sound. For the first ime, scientists have experimentally demonstrated that sound pulses can travel at velocity faster than speed of light. C William Robert's team from Middle Tenessa State University also showed that group of velocity of sound waves can become infinite and negative. Not only that sound can levitate small objects, heat up and also cause fire.

Science works best when in harmony with nature. If we put these two things together, we can discover great technologies that can only come about when consciousness of the planet is ready to embrace them like free energy.

"If you can't explain it simply, you don't understand is well enough"

– Albert Einstien

"It will remain remarkable, in whatever way our future concepts may develop, that the very study of external world led to the conlusion that the content of consciousness is ultimate reality" Eugene P Wigner, a Nobel Prize winner and one of the leading physicists of the 20th century.

"Science cannot solve ultimate mystery of nature because, in the last analysis, we ourself are a part of the mystery that we are trying to solve" Max Plane.

The mysyery and of not understanding how quantum physics work, can onlybe solvedunless physics combines with sprituality. Againunless the subject becones object or vice versa,one cannot Realise Brahman (God) Same way unless we realize that we are made of nothing but quantum energy, for we are a part made by atoms. Understanding quantum physics is nothing but what is already within you is quantum energy. This is to become subject into object. This is the Glory of God. 'Lali dekhan my chali, my bhi hogayi Lal' meaning I went out to see the glory of God, I too became glory.

The Soundarya Lahiri first two stanzas give description of an atom, quantum energy and subatomic particles in a spiritual message – Adi Shankaracharya

English translation:

1. O'Bhagavati! (Goddess) Only if Parma Siva is enjoined with You the Sakti (Energy/consciousness), He is empowered to create. The same Lord is indeed powerless even to move sans Thy Company. O! Mother the celibities such as Hari (Vishnu), Hara

(Siva) and Brahma (the creator) and others, ever worship Thee. In such a context how can one be capable of saluting or praising you without the meritorious effects of (my) yester deed?

As per Hindu mythology an atom compared to Vishnu as proton, Siva as neutron and Brahma as electron. Shakti is the energy which we have been discussing all this time in quantum physics is the Goddess Bhargavi as Shakti/Conciousness, without Shakti (energy/consciousness) matter is inert. This prayer is unifying factor between inert atom and quantum energy.

1. O! Bhagawati (Goddess), Virichi (the Creator) carefully gather minute dust from Thy celebrated lotus feet to accomplish the creativity of all worlds, without any hinderence. Shouri (Vishnu) some how manages to carry this dust on his thousand heads. Hara (Siva) amalgamates this powder and observes the injunction for sprinkling it on Himself as sacred ash (Vibhooti). Here dust means subatomic particles which is gathered by Vishnu and then smeared by Siva by amalgamation.

In atom (which collectively make this Univers) there are three mains ingrediants apart from short lived sub atomic particles-electron, proton and neutron. Lord Shiva (the destroyer) is compared to neutron, Lord Vishnu (Sustainer) to proton and Lord Brahma(creator) to electron. These three combine to form trinity. Shiva causes destruction of universe while Brahma is responsible for creation while Ash is compared to Small short lived sub atomic particles. In nuclear fission explostion it is neuron that causes chain reaction and thus destruction $E=MC^2$. "Pragnanam Braman" Brahman is consciousness/energy. (Ait Up). The followijg say that Brahman alone is the creator of Universe:

Maitreya Upanishad says "In ME the Universe has origin, in Me alone the whole (Universe) subsists, In Me aloneit is lost. This Brahman is Linitless"

"It is Myself" Krishna says in Bha Gita (X.7) "at the end all beings O Son of Kunti, enter into my prakriti and at beginning of a cycle I generate them again"

Conclusion

"Yato va imani bhutani jay ante yena jatani jivani yatprayanyabhisamvi Sati tadvijinasava"

(Thai. Up 3.1)

Meaning That from which all beings are born, having born by which they live. That into which having departed they enter, seek to know That, That is Brahman.

So in the beginning Brahman alone was there, then it desired to become many. And getting heated in austerity it created a pair (Matter and life) which became many.

"tatsatva tadevanuprabisay II tadanupabisya sacca canilayanam ca II Niruktam ca II nilayam canilayalayanamca ca II vignanam cavjnanam ca II Satyam canrtam kina II Tatsatyamityacaksate II

(Thai Up 2.6)

Meaning

Having created entered into it (Universe). Having create He became manifest (asUniverse) and unmanifest, sentient and insentient, real and unreal. The Satya became all that there is. Hence He is called Satya. Satya is Truth.

In finality Brahman alone (God according to Sanatana Dharma ie, Hinduism) **is the Creator of Universe. He exists** in two forms: His subtle form is Nirguna Brahmana ie is unmanifested form and the Manifested form is Saguna Brahman wich is this Universe. Finally I want to end with the following verse:

Ecclesiasts 3:1

"He has made everything beautiful in its time. He has even put eternity in their heart, yet mankind will never find out work that the true God has made from start to finish."

www.ingramcontent.com/pod-product-compliance
Lightning Source LLC
Chambersburg PA
CBHW020625220526
45464CB00001B/30